Gagarin Readings

Unconventional Computing, Space Exploration and Origin of Life

Editor Andrew Adamatzky

Published by Old City Publishing, Inc.
Under the OCP Science imprint,
a member of the Old City Publishing Group

Visit our web site at: oldcitypublishing.com

ISBN: 978-1-933153-46-9

Gagarin Readings

Unconventional Computing, Space Exploration and Origin of Life

Editor Andrew Adamatzky

Preface iii
Andrew Adamatzky

Sensing and Computing with Almost-every-day Materials: Basic Concepts and Case Studies 1
Maria Lis, Tomasz Mazur, Krzysztof Mech, Agnieszka Podborska, Ramesh Sivasamy and Konrad Szaciłowski

How to Build Controllable Biomolecular Circuits Using Protein Filaments 35
Jack A. Tuszynski

On the Positive Role of Noise and Cross-Talk of Elements in Neuromorphic Systems 55
Victor Erokhin

Unconventional Computing in Dendrites – From the Cytoskeleton to the Spines 67
Avner Priel

Interaction of Chemical Reaction and its Environment Enhancing the Advantage of Unconventional Computation 81
Yukio Pegio Gunji

Computing with Autonomous Agents on Graphs 105
Dan V. Nicolau, Jr

Comparison of Effectiveness: Biological vs AI Systems 127
Jordi Vallverdú and Max Talanov

Study of Spontaneous Electrical Low-frequency Oscillations and Morphology of Mycelium Grown in Thermal Polypetide Media 137
Davin Browner, Panagiotis Moukogiannis and Andrew Adamatzky

Kombucha Mats as Responsive Materials 155
Anna Nikolaidou, Panagiotis Mougkogiannis, Alessandro Chiolerio, Mohammad Mahdi Dehshibi and Andrew Adamatzky

Complexity and Nonlinearity of Colloid Electrical Transducers 189
Raphael Fortulan, Noushin Raeisi Kheirabadi, Alessandro Chiolerio and Andrew Adamatzky

Liquid Spaceships for Advanced Autonomous Exploration 203
Alessandro Chiolerio

SPECIAL BOOK EDITION OF PARERS FROM
International Journal of Unconventional Computing
Volume 19, Number 1 and Volume 20, Numbers 1-2, 2025

PREFACE

Gagarin Readings: Unconventional Computing, Space Exploration and Origin of Life

In Spring 2024, to celebrate the science and technology of space exploration, we organized a small but influential workshop titled *"Gagarin Readings: Unconventional Computing, Space Exploration, and the Origin of Life"* in Bristol. The workshop was held on April 11–12, 2024, in honor of the first-ever human spaceflight by Russian cosmonaut Yuri Gagarin. It marked the 63rd anniversary of the flight. Eminent researchers from the UK, Israel, Italy, Poland, Serbia, Canada, Japan, Russia, and Catalonia provided illustrative presentations on their groundbreaking developments in chemical, biological, and astronomical computing. These presentations have been transformed to the book chapters.

The first chapter shows how to sense and compute using materials that are readily available in everyday life. This chapter explores the potential of using commonly available biomaterials for low-energy, easy-to-use electronic devices and sensors, focusing on applications in healthcare, energy, and security. It reviews current biomaterial-based devices and presents a case study on using a special material made from tiny carbon tubes and light-sensitive particles to detect alkali metal ions in water. The second chapter explores the complex relationship between cellular structure and function, focusing on the poorly understood electrical and electromagnetic interactions within and between cells. It discusses the potential to develop an integrated bioelectric model of cells, which could lead to advancements in hybrid technologies, novel bioelectric devices, and applications in nanotechnology and therapeutic modalities.

The third study examines three key properties of neuromorphic systems: the beneficial role of noise in improving classification and forecasting, the importance of element cross-talk for bio-realistic learning and cognitive processes, and the critical need for energy efficiency. It emphasises the necessity of new hardware approaches to address these challenges effectively.

The next chapter explores the electrodynamic properties of dendrites, focusing on the cytoskeleton's role in neural computation. It proposes a model in which microtubules, actin filaments, and their interactions with ion channels and dendritic spines actively contribute to unconventional computing processes within neurons. The fifth chapter presents chemically inspired cellular automata, demonstrating how noise can enhance computational power and break traditional efficiency trade-offs. Another intriguing work introduces a new computing model, "N-systems," based on molecular agents navigating a digraph to solve NP-complete problems, highlighting the potential of massively parallel processing in unconventional computing is presented in the sixth chapter. The comparison between biological systems and digital AI is addressed in a subsequent chapter, illustrating how life forms naturally optimise energy consumption, while modern AI systems fall short in this regard. A further contribution is about the Space Element Reduction Duplication model, examining topological information transmission in a discrete space-time and its implications for understanding physical systems. The book is completed with three chapters. One chapter reports research on the integration of electrical signalling between mycelium and proteinoids in early Earth conditions offers insights into protocognition. The second chapter explores the unique potential of kombucha mats as responsive materials and showcasing the promising future of fungi and kombucha mats as novel, sustainable platforms for intelligent biological systems and advanced materials for soft robotics and biosensors. This follows by the chapter investigating the complexity and nonlinearity of colloidal suspensions when electrically stimulated with fractal, chaotic, and random binary signals. By analysing the responses using entropy, file compression, fractal dimension, and Fisher information measures, the study identifies titanium dioxide as the most complex colloid and highlights the potential of these materials for reservoir computing due to their nonlinear behaviour, the work offers a rapid method for evaluating the suitability of colloids as computational substrates. The last and may be the most provocative chapter explores the potential of liquid cybernetic systems in future space exploration, emphasising the advantages of liquid-based spacecraft over conventional designs. It discusses hypothetical missions involving these innovative systems, highlighting how they could improve the autonomous exploration of distant worlds.

Andrew Adamatzky
Bristol UK

Sensing and Computing with Almost-every-day Materials: Basic Concepts and Case Studies†

MARIA LIS[1], TOMASZ MAZUR[1], KRZYSZTOF MECH[1], AGNIESZKA PODBORSKA[1], RAMESH SIVASAMY[1] AND KONRAD SZACIŁOWSKI[1,2*]

[1]*Academic Centre for Materials and Nanotechnology, AGH University of Krakow, Kawiory 30, 30-055 Kraków, Poland*
[2]*Unconventional Computing Lab, University of the West of England, Bristol BS16 1QY, United Kingdom*

Received: September 1, 2024. Accepted: September 28, 2024.

A major challenge for our uncertain future is to develop such electronic devices and sensors that will be based on commonly available materials, will be easy to use, and will consume little energy. The presented work describes the concepts of using biomaterials for applications in such areas as healthcare and biomedical devices, energy storage and harvesting, or security and authentication. The first part of the presented work contains a literature review of the devices based on biomaterials developed so far. It shows not only the advantages of such devices but also the technological challenges that need to be overcome. The second part describes a system based on cadmium sulphide - multiwalled carbon nanotubes, which can be successfully used to detect alkali metal cations in aqueous solutions.

Keywords: papertronics, biodegradable memristors, reservoir computing, neuromorphic architecture

1 INTRODUCTION

Information technologies are the main driving force of modern economies. Internet of things, wireless digital communication (mobile phone networks,

*Corresponding author email: szacilow@agh.edu.pl
†Authors are listed in alphabetical order.

mobile Internet and other related technologies), GPS, online banking, social media and fast Internet search engines seem to be inevitable tools, which are commonly available to almost everyone. In this context, the problem of digital exclusion (inability to use digital technology) becomes significant and usually touches older generations, irrespective of nationality and economic status. There are only small fractions of society members that give up participation in the digital world in their decision.[1-2] Despite that, the current economy is mainly information-based. Information has become the main good to be traded and one of the most valuable resources. Four of the five largest (by market capacity) companies in the world are based on IT: Microsoft, Apple, NVIDIA, and Google, with the largest petroleum company (Saudi Aramco) in sixth place.[3]

Most of the active participants of the digital (information) world are aware of the real costs of the digital revolution. These costs are, however, very high: both in terms of energy[4-5] and materials.[6-8] Currently, large data centres (usually associated with apparently "free-of-charge" services[9] like Google, Facebook, Twitter, Instagram, PornHub, or Tinder) consume more electric power than large, heavily industrialised countries.[10] The combined energy consumption related with various IT technologies currently constitutes around 10% of world's electricity production and, according to some prognoses, will soon start to dominate over all other consumers of energy, even if the growth of energy production will keep its current rate of increase.[11] Furthermore, the amount of information stored in data centres all over the world is overwhelming and tools for dealing with such a large volume of data are unknown. Constantly increasing amount of data leads to the so-called 'information black hole' – the state in which the amount of available data is too large to be searched and processed. This state has been envisioned by Polish ingenious writer Stanisław Lem in his novel *The Star Diaries*.[12] According to Lem excessive data undergoes a kind of collapse and it is destroyed. In reality, the overwhelming flood of unprocessed data, some of which may be useless, but some may be of crucial importance,[13-15] which from a pragmatic point of view, is equivalent to destruction of the data. This can be mitigated by either changes in data storage habits and policies or by new data management methodologies, including neuromorphic computing.

However, the ultimate performance of computing systems, regardless of their technology, is limited by several fundamental laws related to physical embodiment (representation) of the information, energetic requirement to process, store and transmit.[16-17] The maximal information density is described by the Bekenstein limit, while the data transmission speed is limited by the bandwidth of the transmission channel bandwidth according to the Shannon limit. Other physical limits regarding maximal data processing speed, communication lags, and energy supplies originate from quantum mechanics and the special theory of relativity. The current technological limits create a much stricter bound for the performance of computing systems; therefore, numerous technological obstacles can be overcome in the future. The development of computational performance of a single core is described by Moore's law[18] and Dennard's scaling.[19] Parallelisation of computational tasks is also limited by Amdalh's[20] and Gustafson–Barsis's[21] laws, which define the limits of parallel data processing. Further limitations originate from von Neumann architecture that separates computing from data storage, which implies continuous data transfer between central processing unit and memory, commonly known as von Neumann bottleneck.[22-23] Another significant limitation is thermal management: the current power density at the surface of the microprocessor chip exceeds that of the core of nuclear reactor (i.e. ca. 5 kW/cm^3), this thermal power must be dissipated during the operation of a microprocessor (Figure 1).

Comparison of state-of-the-art microprocessors with energetic performance of a human brain shows that grey matter (with up to 90 billion neurons and 100 trillion synapses) operates at ca. 200 Hz polychronous 'clock' requires only 3 W of power (the total power demand of the brain, including glial cells, is about 20 W), whereas microprocessors comprising 4.5 billion transistors operating at

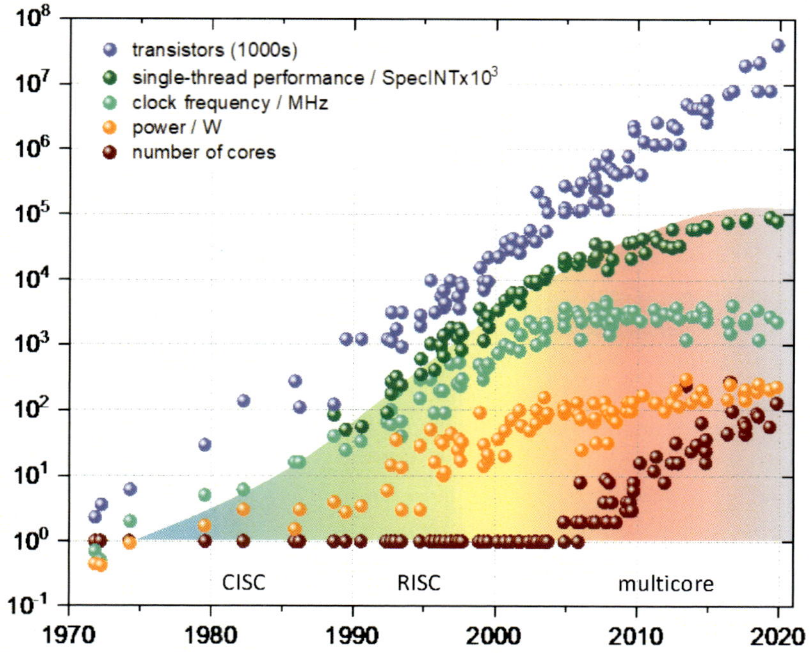

FIGURE 1
K. Rupp's '42 years of microprocessor trend data' graph combined with the Henessy-Patterson graph of computer performance for different processor architectures.[24-25] Raw data courtesy K. Rupp via the github.com repository is available under a Creative Commons Attribution 4.0 International Public Licence.

5.5 GHz require ca. 250 W of power. Comparison of these figures clearly demonstrates the advantage of neural architectures and may serve as an indication of the future development of computing technologies.

Material aspects of computing constitute an interesting field of exploration. According to some philosophical exploration, information is physical and is equivalent to mass. The reasoning is as follows: energy and information are bound by the Laudauer principle,[26] stating the energetic cost of one bit of energy as (1):

$$E = k_B T \ln 2 \tag{1}$$

where k_B is the Bolzmann constant and T is temperature. On the other hand, the mass-energy equivalence is given by Einstein's special theory of relativity (2):[27]

$$E = mc^2 \tag{2}$$

where E stands for energy, m for mass and c is the speed of light. The combination of the two equations leads to an expression showing the equivalence of mass and information (3):

$$m = \frac{k_B T}{c^2} \ln 2 \tag{3}$$

This equivalence, first suggested by Szaciłowski in 2012 as a scientific curiosity,[28] later became a subject of a serious debate.[29-32] Despite serious arguments by Laudauer and Wheeler[33-34] information should be rather understood as a feature of matter than an individual entity.[32] In this context any transformation of matter cannot be considered as immediate information processing; however, it does not exclude application of various molecular phenomena in computing.[28, 35-37]

Therefore, the future development of computational techniques, circumventing the problems of energetic efficiency, information black hole, and shortage of rare elements, will require not only new computational paradigms (neuromorphic computing, physical reservoir computing, computing in memory) but also new materials suitable for fabrication of neuromorphic architectures. Taking into account limited supplies of rare elements on Earth, the quest for the application of Earth-abundant elements in computing devices is of utmost importance.

Here we present an overview of various computing applications based on ubiquitous materials: biodegradable memristors for neuromorphic information processing, paper-based electronics (papertronics), and a simple nanomaterial-based sensor, which despite chemical simplicity offers complex analysis of ionic media and also allows implementation of optoelectronic logic gates and possibly synaptic devices.

2 BIODEGRADABLE MEMRISTORS FOR NEUROMORPHIC INFORMATION PROCESSING

As was stated in the introductory section, information processing is possible in almost every known type of material – there have been multiple reports on that matter. However, nowadays only a small percentage of materials are widely used, inorganic layers or hybrid organic-inorganic occupying the mainstream hardware. Such a situation leaves many possible good computational materials unexplored. Meanwhile, a variety of simple materials in their structure, or systems of such are readily available, and with minimum effort can contribute as active layers in processing test devices – for now only in laboratory scale. Such a strategy can have many positive possible outcomes - limiting digital exclusion being one of them - as cheap and reliable electronics contribute directly to social prosperity.

The vast amount of "biomaterials" serve as nonvolatile memory or resistive switching devices.[38] Contrary to typical inorganic or hybrid semiconductive materials, they are widely available, sustainable, and by definition biocompatible. Usually that means low or lack of toxicity, and in addition their durability and degree of planned degradation can be easily tuned, e.g. by controlled exposure to humidity/water. Naturally, as a subset of organic compounds, they feature a non-rigid structure, which leads to their implementation in flexible electronic hardware. In examples presented below, the device performance is comparable with inorganic/hybrid ones – they are characterised by low power consumption, long retention times, proper cycle endurance, and operate within CMOS voltage limits. All of the above make biodegradable and biocompatible resistive switching devices a real alternative to currently implemented systems.

Another technological challenge to face while fabrication devices is the complexity of the entire process. Commercially available electronic consumables are mostly multilayered microchips in a complementary metal-oxide-semiconductor technology. The latter requires specialised photolithographic equipment, elevated temperature, and high vacuum and generates a lot of chemical waste. The technological shift to biodegradable materials would simplify processing and limit the use of organic solvents. Most simple operational devices can be fabricated in an MIM architecture (metal-insulator-metal) or more general top-electron-functional material-bottom electrode. When using a device entirely, electrode materials must also be chosen, taking into account their compatibility with the functional layer, as well as possible mechanisms they could introduce to the operation of the device. Typically, they are divided into categories: silicon-based, nitrite-based, alloy-based, oxide-based (ITO, GO), and pure metal-based. The last two are the most widely used.

So far, there exists no formal division of biocompatible materials, yet based on literature, several groups can be distinguished, with at least several examples in each categories.[38] Distinguishable classes of materials are in this case proteins, carbohydrates, nucleic acids, or their mixtures. The additional class also includes natural mixtures, such as edible products.

Protein-based devices have been presented so far in many instances. The most simple one, was constructed of the egg-based layered memristive devices.[39] The device was manufactured following MIM architecture and experienced measurable resistive switching phenomena. The performance of the device was modified by means of post-synthetic thermal treatment, leading to redistribution of oxygen diffusion paths and increasing the ON-OFF ratio. The intrinsic filament formation/rupture cycling was metal-based, Fe from other proteins present in egg white ovotransferrin. Egg white is also a source of charge carrier traps, amino acids, which also contribute to the overall layer conductivity. Interestingly, these devices are characterised by cycle endurance of the order of up to 10^4, high switching ratio and it supports several hundreds of voltage scan cycles. These properties make them potentially operational in real-life tasks, with limitations mentioned before. The whole egg-based group of memristive devices,[40-42] despite having proper operational parameters, rather stays as scientific curiosities and serves as proof-of-concepts, leaving their applications to very specific tasks.

Protein-based resistive switching devices operate also on protein mixtures: collagen, ferritin, lignin, or even silk. These designs will mostly find their usability in applications that require bioresorbable and nontoxicity - i.e. one time usage devices. The addition of flexibility makes them a good choice for medical applications. Also, it is much easier to produce transparent or semitransparent devices. All of the above are in previously described MIM configurations, operating typically within -5V to +5V, with the dominance of space-charge-limited conductivity (SCLS) conductivity mechanism, laying beneath resistive switching phenomena.

The second main group of biodegradable processing devices is carbohydrate based, including monosaccharides, polysaccharides, cellulose, chitosan, sugar, glucose, starch. Resistive switching properties arise mainly from the presence of hydroxide ions. Examples of such systems are made out of cellulose, polysaccharide present in many plants, bacteria, and algae. Cellulose nanofiber paper was used as a functional layer in the Ag/CNP/Pt (either on silicon or elastic substrate) device,[43] yielding low threshold operations (between 0.2V and -0.3V), 10^5 s state retention time, and being able to operate upon bending.

Among polysaccharides, chitosan, composed of randomly distributed deacetylated units of glucosamine, is another material suitable for information processing devices. In addition to being easy to dissolve, it has antimicrobial properties: thin films are easy to fabricate and biodegradable. In addition, it introduces transparency and flexibility to the potential device. The sample system has the configuration of Mg/Ag-doped chitosan/Mg (on elastic substrate),[44] operates between 2V and -3V, retains operational character upon 50 voltage sweeps and 10^4 bending cycles. After use, it can be easily discarded - simply putting it into water, where it is dissolved within 10 minutes.

The third main group of biomaterials incorporated in pioneer resistivity switching devices is nucleic acids and their polymers (DNA). Biomacromolecules, constituting functional layers, are determined for simple processing, prolonged stability, biocompatibility, and self-assembling properties. In addition, they promote nanoparticle use, being efficient matrices for them, which allows for fine tuning of electric conductivity. The sample system was constructed in the form of Ag/(DNA)–cetyltrimethylammonium (CTA) AgNP/ITO.[45] The device operated on an SCLC principle and a registered retention time of 10^5 s. The addition of Ag nanoparticles was crucial for the sufficient amount of charge carrier to flow through the layer.

The layered device architecture does not define types of matching materials, and rather acts as a platform which allows for even more surprising material choices. The constitution of the functional layer of devices does not need to be precisely known as long as the desired resistive switching phenomena are present. What follows into that category (fourth one) are edible products, with natural organic honey being one of them.[46] The said material serves as a resistive switching memristive layer, suitable for data storage, transmission, and other memory tasks. Still, the system incorporates ITO as the bottom electrode and sputtered Al as the top electrode. The switching mechanism is based on the principle of both metal ion movement through the layer, as well as the existence of trap centres: OH groups (deep traps) and structural defects (shallow traps). The system itself did not only present the hysteresis curve, but was also responsive to typical plasticity traits – such as PPF (pulsed pulse facilitation), where the second electric pulse, appearing after short time from the first one generates additional current, due to the formed additional conductive paths between electrodes.

Another example of a device based on edible mixture could be whisky-based resistive switching device.[47] Evaporated whisky droplets yielded in complex layers containing thousands of organic components, sandwiched between metallic electrodes (Ag and Co). The authors underline the fact that functional devices can be made under ambient conditions, with no corrosive chemicals. The operational system was classified as electrochemical metallisation memory (ECM), with metallic filament (Ag) grown/ruptured across the functional layer. The device is operational between -1.5V to 2V, with on/off ratio in the order of 10^2 - 10^3, and can sustain tens of switching cycles.

The final example of the preparation of biocompatible memristive devices is the direct incorporation of leaf extract of the traditional ayurvedic medicinal herb called giloy (*Tinospora cordifolia*).[48] The systems, deposited on the FTO surface, with a top metallic electrode (Ag), show reversible bipolar and non-volatile switching behaviours. In addition, the film composite with quantum carbon dots introduces photo tunable resistive switching characteristics. The additive, carbon quantum dots, can also be prepared according to sustainable and simplicity strategies, e.g., on a hotplate, in the carbonisation process of lemon juice.[49] Microwave-assisted synthesis of these nanoparticles is also possible-directly from plant extracts – yet this approach requires more advanced laboratory equipment park. Typically, the underlying mechanism of resistive switching is a combination of the formation/rupture of metallic (Ag) filament, joint with ionised molecular vacancy interlayer movements.

Having been nothing more than a bit of scientific curiosity for now, biocompatible memristive systems show that even with the scarcity of resources it is possible to construct fully operational electronic devices. A short compilation of the described examples is presented in Table 1. The devices in

Device Configuration	Mechanism of Switching	ON/ OFF Ratio	Operational Voltage Tanges	State Retention Time (s)	Cycle Endurance	Ref.
Al/albumen/ITO	SCLC	10^3	-3V to 4V	10^4	over 500	39
Ag/cellulose nanofiber paper/Pt	Ag conductive filaments	10^6	-0.5V to 0.5V	10^6	100	43
Mg/Ag-doped chitosan/Mg/elastic substrate	SCLC	10^2	-3V to 2V	10^4	60	44
Ag/DNA-CTA-AgNP/ITO	SCLC	~200	-6V to 6V	10^5	-	45
Al/honey/ITO (treatment: 90*C for 8h)	Al filament, deep and shallow electron traps	10^2-10^3	-3V to 3V	10^4 (estimated 10 years)	20	46
Al/whisky deposit/ Co	Ag filament	10^2	-1.5V to 2V	600	20	47
FTO/plant extract/ Ag		10	-3V to 2V	10^3 (estimated 10 years)	1000 switching cycles. 100 mechanical flexing cycles	48

TABLE 1
Exemplary processing systems made of biodegradable materials

description operate on resistive switching principles and thus are able to utilise the non-von Neumann concept of information processing. That architecture allows simplification of the device preparation stage as even one functional layer is enough, but also secures low power consumption, leaving high speed computations readily available. The potential disadvantage of presented systems is the fact that they share the hardware platform with modern electronics. In other words, electrode layers still require state-of-the-art sputtering or ion etching systems. In order to completely move towards sustainable and easy-to-prepare devices, development of completely new hardware platforms is crucial. In this regard, the papetronics concept of electronic devices seems to be of much interest, and will be presented in the next section.

3 PAPERTRONICS: AN ADVANCING ELECTRONIC COMPONENT

Can we blend paper with modern electronics? This question drives us to explore the new materials. Papertronics is an expanding field that combines traditional materials with cutting-edge technology for practical applications.

Primarily composed of polymeric cellulose, the paper has been utilised for more than two thousand years due to its distinctive characteristics, which include an interlocking 2D fibrous structure ranging from millimetre to nano scale.[50] This intricate mesh not only provides an excellent mechanical strength but also facilitates the absorption and manipulation of fluids, making paper an ideal perfect material for electrochemo-optomechanical sensors and devices. Paper exhibits adjustable stress-strain relationships, allowing it to possess either a softness comparable to that of biological tissue or a hardness comparable to that of Kevlar, with a potential elastic modulus for cellulose nanocrystals. Cellulosic fibres can be used in conjunction with metallisation, conductive coatings, nanotubes, and graphene to create patterned electrical characteristics. Paper also has an abundance of materials, ease of handling and processing, excellent biodegradability, and biocompatibility, suitability. The roll-to-roll manufacturing process for paper can exceed speeds of 100 km h^{-1}. It is being widely used as a flexible substrate, and cost-effective compared to polyethylene terephthalate and polyimide.

The concept of printing electronics on flexible paper and plastic substrates has a long history. In the late 1960s, Brody and Page experimented with a stencilling approach to deposit thin-film transistors onto paper substrates. However, their attempt to develop flexible electronics on paper faced commercial failure due to the required vacuum, which hindered cost-effectiveness and fragility of the materials. Nevertheless, 40 years later their techniques found application in flexible electronics such as ATM cards, medical sensors and many others. The challenge of achieving low-cost production remained a persistent issue. Recently, paper-

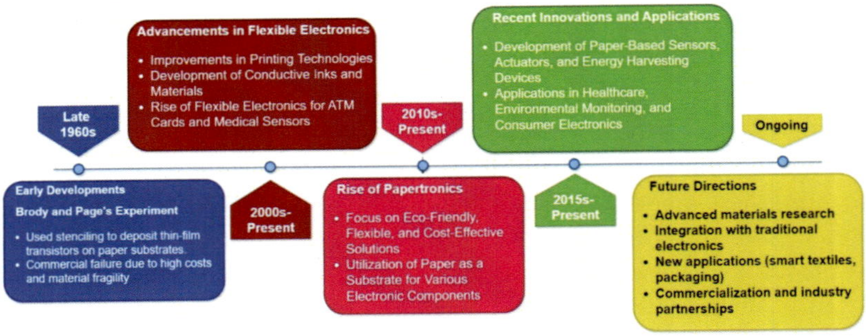

FIGURE 2
Roadmap of the flexible electronics research

tronics represents a pioneering approach that uses the paper substrates, leveraging their affordability, flexibility, and eco-friendly disposability. This harmonious fusion is driving the development of novel devices capable of interactive experiences, advanced sensing capabilities, and unique functionalities.

Recent advances in the field of paper electronics have been significant. Numerous examples highlight the innovative applications and progress in this area. In this chapter, we discuss the architectures of papertronics, fabrication techniques, advantages, challenges, application, and future directions.

3.1 Papertronics Architectures

In terms of electronic devices on paper substrates, the field of papertronics combines a variety of scientific and engineering disciplines, such as fluidics, electronics, photonics, mechanics, and chemistry. The growing interest and advancements in this field of device fabrication have been reflected in the substantial increase in research and publications on paper electronics over the past decade. Papertronics devices capitalise on the distinctive characteristics of cellulose-based paper, including its renewability, flexibility, and fibrous structure, to develop innovative and multifunctional electronic systems. The overall functionality and performance of the devices can be attributed to the breakdown of this architecture into numerous essential components and layers.

Each integrated layer in papertronics serves a unique purpose. The substrate layer at the base is typically composed of cellulose-based paper, functionalised paper or modified Paper, which offers a renewable and flexible support.[50] Electrical channels are formed by the conductive layer, which is often made of silver nanoparticle ink,[51] conductive polymers such as polyaniline or poly (3,4-ethylenedioxythiophene) (PEDOT).[52-53] Proper insulation between conductive channels is ensured by the insulating layer, which uses dielectric materials such as polyvinyl alcohol (PVA)[54] or

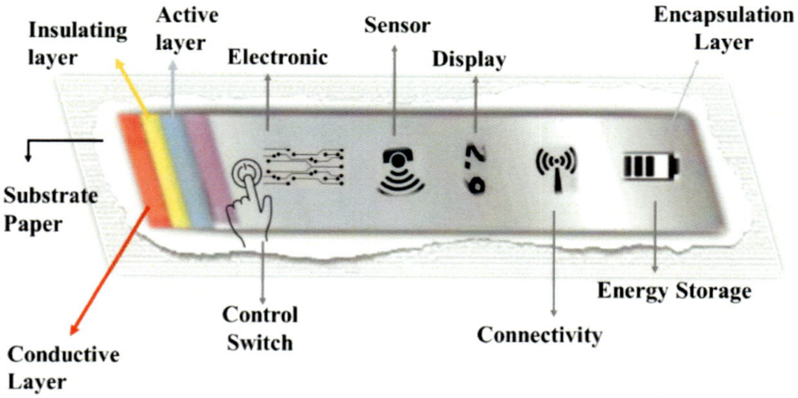

FIGURE 3
Architectures and applications of Papertronics

poly(styrene sulfonic acid) (PSS).[55] The active layer comprises organic semiconductors such as poly(3-hexylthiophene) (P3HT) [55] and poly(9-vinylcarbazole) (PUK),[56] which facilitate electronic operations such as sensing and switching. The metal coating layer and polyethylene terephthalate encapsulation layer.[57] This multilayered approach allows for the creation of sophisticated, flexible, and eco-friendly electronic devices suitable for various applications.[50, 58-59]

3.2 Device Fabrication Techniques

Significant advances have been made in the technology of papertronics, which involves integrating electronic functionalities into paper. Paper with diffusion, printing, and stacking capabilities facilitated the reduction of the complexity of production steps. Various fabrication methods such as printing, coating, lithography, laser processing, embedding, chemical process, and laminations are implemented to produce papertronic devices for a multitude of applications. Consolidate synthesis techniques are represented in Figure 4.

3.2.1 Printing Process

The process of printing is one that is widely used and can be applied to a wide variety of substrates and ink types. There are a few different printing processes that can print inexpensive circuits.[60-61]

(a) Screen printing: This is a commonly used technique in Papertronics because it is straightforward and effective. This method entails the application of conductive ink by exerting pressure through a mesh screen onto the paper substrate, resulting in the formation of electrical circuits and patterns. Screen printing is very advantageous for the efficient production of large-scale electronics.[62]

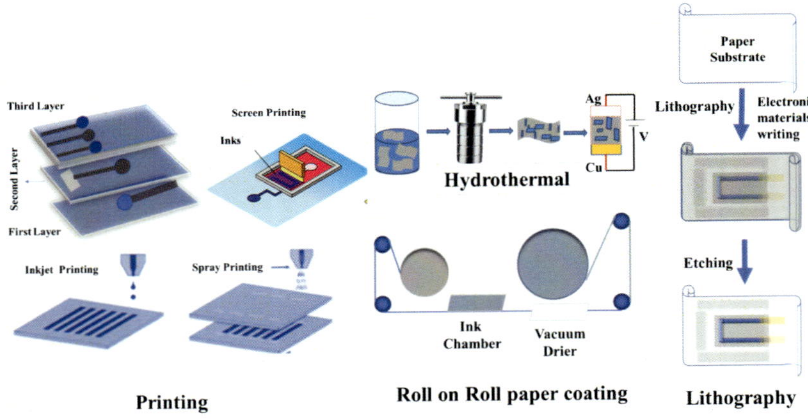

FIGURE 4
Consolidated Papertronics device fabrication techniques.

(b) Inkjet printing: This enables precise use of conductive inks on paper, resulting in the creation of intricate designs. The technique offers significant benefits for producing complicated and elaborate designs, which makes it well suited for sophisticated electronic circuits.[53]

(c) Spray printing: The ink turns into tiny particles and then applied to the paper surface using a specialised nozzle. The spray can be manipulated to attain consistent distribution and accurate formation. This printing technique provides numerous benefits, such as rapid coverage of wide surfaces, compatibility with diverse ink substances, and the possibility of efficient mass production of sensors, display, and energy devices.[63]

3.2.2 Coating

The process of coatings requires the deposition of thin layers of conductive material, such as silver or graphene, onto paper substrates. These coatings establish the electrical pathways essential for the operation of electronic devices. Some organic semiconductors or piezoelectric coatings are used to provide specialised qualities, such as sensing capacities or actuation.

3.2.3 Lithography

It is an essential method for creating precise electronic circuits on paper substrates. Photolithography is a process that uses light-sensitive materials and masks to generate accurate patterns. This method is crucial for precisely establishing the complex patterns needed in electrical devices. After the patterning process, etching is used to selectively remove undesired material from the paper substrates, thus precisely outlining the electronic structures. However, this process is critical for producing intricate features and patterns that are necessary for the proper functioning of the paper-based electronic devices.[64]

3.2.4 Hydrothermal

Few reports on hydrothermal synthesis of paper mesmeriser were reported in which the first step is, cellulose paper was dipped in a solution containing 3 mmol of ammonium metavanadate (NH_4VO_3) for 1 hour, then dried at 70°C for 15 minutes. The coated paper was placed in an autoclave with a solution of 30 mmol NH_4VO_3 and 20 mmol hydroxylamine hydrochloride ($NH_2OH{\cdot}HCl$) and heated at 200°C for 20 hours. After the hydrothermal process, the paper was dried again at 70 °C for 15 minutes. Finally, silver paste was applied as contacts on both ends of the V_2O_5-coated cellulose paper.[65] similarly Parikshit et al.[66] synthesised MoSSe, $ReSe_2$, and SnS_2/Gr on cellulose paper-based sensors.[66]

3.3 Applications of papertronics

Electronics typically categorise components into two primary classifications: passive and active. Passive components consist of resistors, capacitors, and inductors, whereas active components cover transistors, operational amplifiers, displays, and batteries. When considering paper-based electronics, passive components are generally simple to manufacture, offer a viable solution to the growing problem of electronic waste. As a result of its insulating characteristics, paper has been used in the construction of many active components. The rise of papertronics has garnered significant attention in this pioneering field. This includes the development of photodetectors,[58] supercapacitors [52], sensors,[58] transistors, memory, sensors,[67-68] and piezoelectric sensors.[58]

3.3.1 Consumer electronic components

The resistor was created on a 3 MM Whatman filter paper by filling it with conductive ink that can be adjusted to control the resistance. The paper fibres have inherent electrical insulating qualities. The ink that can modulate conductivity, the paper becomes functionalised, and it can work as an adjustable resistor for electrical currents. The process enables the production of paper-based resistors by utilising microfluidic methods to introduce ink, which easily spreads through the paper via capillary action without requiring external energy. Using this innovation, the precise resistance values can be adjusted with precision by manipulating the conductivity of the ink and the quantity deposited. Poly(3,4-ethylenedioxythiophene) polystyrene sulfonate (PEDOT:PSS) is used as a conductive media.

The capacitor is a crucial passive electronic component in integrated systems. It plays a key role in signal processing, calibration, timing, and filtering, contributing to the stability and performance of electronic devices. Capacitors help suppress transients, condition signals, and support the creation of oscillatory circuits. As with the resistors, PEDOT:PSS was chosen for the top and bottom electrodes of the capacitor because of its excellent pseudocapacitive properties. To improve the electrical conductivity of these electrodes, dimethylsulfoxide (DMSO) was added to the PEDOT:PSS mixture. Poly(vinyl alcohol) (PVA) gel with phosphoric acid (H_3PO_4) electrolyte was used as a dielectric layer. The research reported here has successfully produced capacitors with outstanding tunability, demonstrating capacitance values ranging from 10 to 190 μF.[69]

3.3.2 Healthcare and biomedical devices

Yetisen et al.[70] work highlights the noteworthy uses of paper-based microfluidic point-of-care diagnostic devices with a particular focus on their potential in healthcare applications. Paper-based diagnostic equipment, such as wearable health monitors and lateral flow tests, provides an economical and disposable alternative for point-of-care diagnostics. These devices have the ability to track and measure a range of health indicators, such as glucose levels, pH, and biomarkers associated with diseases.[70] A recent study conducted by Martinez et al. revealed that camera phones and paper-based microfluidic devices are used for real-time, off-site diagnosis in the field of telemedicine, especially in underdeveloped nations.[71]

Another report on growing concern about the potential contamination of canned food due to packing materials, namely heavy metals such as tin (Sn(II)) and lead (Pb(II)). These pollutants can lead to a range of health concerns, such as gastrointestinal disorders, cardiovascular and neurological consequences, and potentially even cancer. There is an increasing demand for portable, simple instruments to rapidly identify these metals. An effective approach is the utilisation of a paper-based sensor device (PAD), which is simple to fabricate, economical, and designed for single use. This technology utilises a combination of BiNP (bismuth nanoparticle), graphene, and Ag (silver) to detect both Sn(II) (tin) and Pb(II) (lead) simultaneously in canned food. The method provides excellent selectivity, sensitivity, stability, and repeatability. The results can be obtained in a few 2 minutes using only a 100-μL sample.[72]

3.3.3 Energy storage and harvesting

In addition, applications that involve energy storage and harvesting can benefit from breakthroughs made by papertronics. These innovations include paper-based batteries and flexible solar cells. According to Zhu et al.,[73] these devices take advantage of the adaptability and biodegradability of paper to provide sustainable and adaptive. Papertronics is a company that helps facilitate the development of interactive displays and smart packaging in the market for consumer electronics. According to Xiong et al.,[74] these applications make use of the low-cost and environmental benefits of paper to develop innovative user interfaces and interactive features.

Another successful development of a paper-based microbial fuel cell (MFC) that has been pre-inoculated with lyophilised cells. This design enables a long shelf life for the battery and allows for activation with saliva, making it suitable for self-powered applications in resource-limited environments. The MFC was fully integrated into the paper substrate, with all functional layers created using precisely patterned asymmetrical wax. Conductive paths and microfluidic channels were also incorporated for the anodic reservoir. The MFC device is fabricated by first preparing conductive paper treated with PEDOT combined with 5 wt% DMSO and 2 wt% 3-glycidoxypropyltrimethoxysilane. For air cathodes, activated carbon is used with a nickel spray and 5 wt% Nafion solution to enhance conductivity. Electrical contacts are established using conductive silver paste to ensure effective connectivity throughout the device.

The intrinsic capillary forces and the engineered reservoir facilitated rapid and effective adhesion of microbial cells to the conductive paper matrix, leading to immediate and robust power generation, even with microlitre-sized samples of activating fluid. By connecting sixteen MFC units in series on a single sheet of paper, the system could power an LED using artificial saliva samples. If validated with actual body fluids, this innovative MFC could lead to the creation of a self-powered point of care (POC) testing device, enabling patients to perform self-testing and manage their care in various settings, including at home, in emergencies and on the battlefield.[69]

3.3.4 Security and authentication

Security and authentication systems are another crucial sector in which papertronics has a significant impact. The use of this technology allows for the identification of individuals based on their facial characteristics as well as the identification of genuine banknotes.[58] Moreover, the technology of papertronics is used in the biometric verification process for paper-based devices to improve security controls. To maintain secure accessibility and identification, this technology provides a solution that is both low-cost and customisable, as stated by Jang et al.[75]

In addition to the aforementioned uses, Papertronics sensors and devices are also being used in various other domains. They are commonly employed as chemical sensors, optical sensors, food safety sensors, wearable healthcare sensors, and electrochemical sensors, serving a diverse array of applications.

3.4 Current Challenges

Fabricating electronic components on paper surfaces presents unique challenges, particularly switching, as conventional silicon production techniques are impractical. Switching from a printed circuit board (PCB) to a paper substrate requires the adjustment of materials. The main problem encountered was the intrinsic roughness of a porous material. In this instance, innovative

Applications	Fabrication Techniques	Paper Substrate	Electronic Active Components	Ref.
Transistors	Printing	bond, vegetal, and Lumi Silk,	PEDOT: PSS, carbon electrodes, cellulose-based electrolyte	53
Memristor	Coating	Cellulose paper	Cu/Gr/MoS_2/Ag.	61
Sensors	Pencil drawing Handwritten	Cellulose paper Cellulose paper	Pencil MoS_2/paper-based sensor.	68 67
Batteries	Paper devices	CNT-cellulose-RTIL sheets	Li batteries and hybrids	76
Supercapacitor	Printing	Wax infused	PEDOT: PSS ink n-rGO	77
Resistors	Patterned	Hydrophilic paper	Pd-CNT-rGO/PEDOT: PSS ink	77
Antenna	Inkjet Printed	hydrothermal method	SnS2 on Gr/cellulose paper	78

TABLE 2
Summary of selected applications paper-based electronic in the literature

techniques to achieve a smooth paper surface or to produce papers with several layers and porous centres while maintaining smooth surfaces.

The second challenge in the development of paper electronics was the high cost and limited availability of suitable inks. Research is essential for the development of new inks that include unique qualities, such as semiconducting capabilities, while also being cost-effective.

The third challenge is to identify the problems related to devices and packaging. The low heat resistance prohibits the use of wet and high-temperature processes such as the sol-gel method, chemical vapour deposition (CVD), and ion implantation, essential for conventional semiconductor production. These techniques typically require temperatures beyond what the paper can withstand. Due to the thermal instability, direct application of these technologies is not feasible, necessitating alternative approaches. For instance, the sol-gel process involves the transformation of a liquid solution into a solid gel, typically requiring wet conditions and high temperatures to achieve desired material characteristics and conditions unsuitable for paper. CVD, where gaseous reactants form solid materials on a substrate, also requires elevated temperatures for effective chemical reactions and material deposition. Ion implantation, involving high temperatures to modify substrate properties through ion bombardment, is also incompatible with paper. Given the potential impact of humidity on paper morphology and performance, it is necessary to provide extra packaging for all paper-based devices to maintain their quality and shelf life. In addition, the paper may undergo biodegradation as a result of the enzymatic activities of various microbes to which it is exposed.

Therefore, it is crucial to develop materials and manufacturing methods tailored to the limitations of paper. Investigating fabrication methods and materials that operate at lower temperatures is essential. Techniques such as inkjet and screen printing, along with other additive manufacturing technologies, are better suited for paper substrates. These methods can be refined to produce functional electronic components without compromising the integrity of the paper substrate.

3.5 Future directions

Papertronics is on the verge of revolution in numerous fields. Enhancing the qualities of materials, such as the advancement of conductive materials such as graphene and biodegradable polymers, will improve performance and sustainability. Enhanced manufacturing methods, such as high-resolution printing and scalable roll-to-roll production, will enhance the accuracy and cost efficiency of paper-based devices. Integrating with Internet of Things (IoT) ecosystems will allow the use of inexpensive, disposable sensors and intelligent packaging solutions. In addition, the advancement of healthcare applications will result in the development of advanced wearable health monitors and diagnostics that can be used at the point of treatment. Progress in energy harvesting and storage, such as flexible photovoltaic cells and batteries with large storage capacities, will provide the necessary power for increasingly complex gadgets. Additionally, the advancement of interactive displays and touch/gesture recognition technologies will improve user interfaces. Furthermore, the implementation of advanced security features will address counterfeiting issues and guarantee secure communication. These advances will propel the development of papertronics, providing environmentally friendly, adaptable, and economically efficient solutions for various uses.

4 CASE STUDY: SENSING AND COMPUTING WITH NANOCOMPOSITES: CDS-MWCNT

The examples discussed so far demonstrate that the large diversity of materials can be used for sensing and computing. It is in line with the concept that even a rock can compute,[79] which has also been recently demonstrated.[80-81] Here we show, how relatively simple materials (multiwalled carbon nanotubes and cadmium sulphide nanoparticles) can be used as a sensor yielding a complex response on illumination in various ionic environments.

4.1 Experimental Details

All reagents and solvents were supplied by Sigma-Aldrich.

4.1.1 Synthesis of cadmium sulphide nanoparticles

1 gram of cadmium acetate dihydrate was placed in a round bottom flask and dissolved in 30 cm^3 of ethylene glycol. 286 milligrams of thiourea dissolved in 30 ml of ethylene glycol were added with vigorous stirring. Finally, 5 ml of morpholine was added to the mixture. The flask was equipped with a reflux condenser and placed in a microwave reactor operating at 800 W for 1 minute and allowed to cool for approximately 30 minutes. The dark yellow product was isolated by centrifugation at 800 rpm, washed several times with deionised water and air dried.

4.1.1.1 Synthesis of crown-ether-modified carbon nanotubes

4.3 mg of multiwalled carbon nanotubes and 4.3 mg of dibenzo-18-crown-8 (D18C6) were placed in a beaker containing 10 cm^3 of acetone. The resulting mixture was sonicated for 15 minutes. Finally, the homogeneous suspension was poured into a round bottom flask (100 ml) and acetone was evaporated in a water bath.

4.1.2 Synthesis of CdS-nanotubes composites

Cadmium sulphide was synthesised in the presence of either unmodified or modified multi-walled carbon nanotubes (4 mg) as described above.

4.1.3 Instrumentation

Transmission electron images have been recorded on the Tecnai TF 20 X-TWIN instrument (FEI, USA) at 200 kV acceleration voltage.

The Mott-Schottky analysis was performed using Biologic SP-200 potentiostat/galvanostat in a standard three-electrode system consisting of the Ag/AgCl (3.5 M KCl) reference electrode, the Pt wire gauze counter electrode, and (CdS, CdS-MWCNT and Cd-MWCNT-D18C6)@ITO working electrodes (material deposited on the ITO substrate). Each electrode was kept at the corresponding potential for 5 seconds prior to measurement. The data points for the selected potentials were recorded in the dark at a frequency of 1 kHz with a modulation signal amplitude of 15 mV. Measurements were made in electrolytes containing 0.1 M of $LiNO_3$, $NaNO_3$, KNO_3, $RbNO_3$, and $CsNO_3$, respectively. Electrolytes were purged with Ar for 20 min directly before measurements.

Photocurrent measurements for all materials were provided using a photoelectric spectrometer (Instytut Fotonowy, Poland) in a three-electrode system with a working electrode: ITO-coated foil with deposited materials, an Ag/AgCl electrode, as the reference electrode and a platinum counter electrode. Measurements were carried out in 0.1 M aqueous solutions of appropriate alkali metals ($LiNO_3$, $NaNO_3$, KNO_3, $RbNO_3$, $CsNO_3$) in the potential range from -200 mV to 400 mV, illuminating the system with light (wavelength range 300 nm - 800 nm) which were applied to and constituted a working electrode (CdS-MWCNT-D18C6@ITO)

4.1.4 Quantum-chemical modelling

Quantum-chemical modelling was performed using the Gaussian 09 E.01 software package at the cam-b3lyp/sto-3g/def2tzv level of theory. The calculations used a carbon nanotube model containing 280 carbon atoms. Molecular geometry, charge distribution and frontier orbitals, and electrostatic potential isosurfaces were calculated using GaussView 5.0. Calculations were performed for systems composed of a carbon nanotube and a D18C6 molecule and a cadmium sulphide molecule. At a later stage, the system was expanded with alkali-metal cations (Li^+, Na^+, K^+, Rb^+, Cs^+) in order to check the interactions between the cation and the hybrid material.

5 RESULTS

Solvothermal decomposition of thiourea in the presence of morpholine and cadmium acetate results in the formation of morpholine-stabilised cadmium sulphide nanoparticles (Figure 5a). If the reaction is performed in the presence of multi-walled carbon nanotubes, the growth occurs mostly at the surface of the nanotubes, resulting in a composite material with a good electrical contact between the components (Figure 5b). In these materials, the surface of carbon nanotubes is decorated by dibenzo-18-crown-6 molecules bound via π-stacking interactions (Figure 5c). The surface of CdS nanoparticles is decorated with morpholine molecules coordinated to the cadmium centres on the surface via nitrogen atoms (Figure 5d).

Because of the presence of two ethers (morpholine and crown ether), all studied materials have a weaker or stronger tendency to interact with alkali metal cations: the oxygen atom in the morpholine molecule is a weak ligand for alkali metal cations, but its presence at the surface of CdS may result in a local increase in cation concentration. On the other hand, dibenzo-18-crown-6

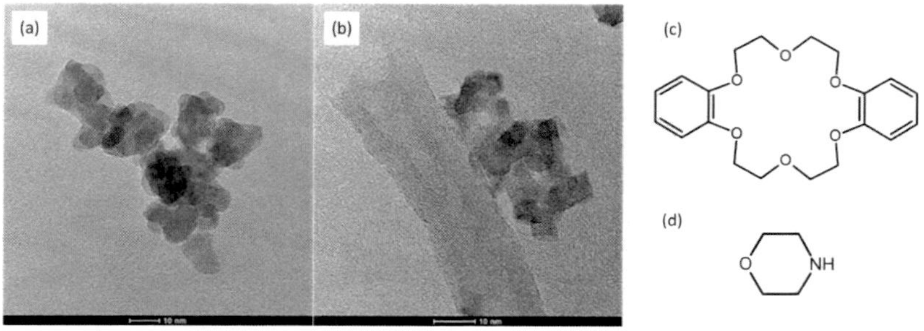

FIGURE 5
Transmission electron microscopy images of CdS (a) and CdS-MWCNT (b) materials. Molecular structure of dibenzo-18-crown-6 receptor molecule (c) and morpholine ligand (d).

is a specific macrocyclic receptor for alkali metal cations, especially potassium; however, interaction with other cations is also significant.[82-85] As these interactions should affect the charge density at the surface of semiconducting nanoparticles, the Mott-Schottky (M-S) analysis seems to be the most logical method to study these interactions.

The slope of the Mott-Schottky plots indicates that all CdS-based materials exhibit *n*-type conductivity. Determining the imaginary part of impedance (Z_{im}) enables the evaluation of the capacitance of the space charge region (C_{sc}) expressed by the following equation (4):

$$C_{SC} = \frac{-i}{2\,\pi f Z_{im}} \tag{4}$$

where i is the imaginary unit and f is the frequency of the ac-voltage signal. The Mott-Schottky approximation of the capacitance-voltage dependency for *n*-type semiconductors is given by (5):

$$\frac{1}{C_{SC}} = \frac{2}{\varepsilon\varepsilon_0 e N_D A^2}\left(E - E_{FB} - \frac{kT}{e}\right) \tag{5}$$

where ε is the dielectric constant of the material, ε_0 is the vacuum permittivity ($8.85\cdot10^{-12}$ F/m), e is the elementary charge ($1.602\cdot10^{-19}$ C), N_D is the electron donor concentration (cm^{-3}), A is the surface area, E is the applied potential, k_B is the Boltzmann constant (1.38×10^{-23} J/K) and T is the temperature on the absolute scale. The donor density and the flat band potential can be determined on the slope of the Mott-Schottky plot and the extrapolation of the dependency to C=0, respectively.

Figures 6 and 7 present ion- and material-dependent representations of the same collected Mott-Schottky characteristics recorded for all prepared sensing electrodes. Results were presented in two ways to reflect the differences in Mott-Schottky characteristics resulting from changes in the composition of materials in the presence of selected ions (Figure 6) and the differences in Mott-Schottky characteristics corresponding to changes in the capacitance of the space charge region resulting from the presence of particular ions of specific physicochemical properties (i.e., ionic radius, ionic mobility, etc.) (Figure 7).

Figures 6 and 7 do not reveal any clear effect of sensing materials or modifying the environment composition on observed E_{FB} values and the Mott-Schottky slope. The most noticeable changes were indicated for the ternary material (CdS-MWCNT-D18C6) sensing material in the presence of Li^+, Na^+, K^+, and Cs^+. The C_{sc} values were lower than those observed for the remaining materials in the investigated potential range. The changes in material-electrolyte (Li^+, Na^+) interface properties resulting from the modi-

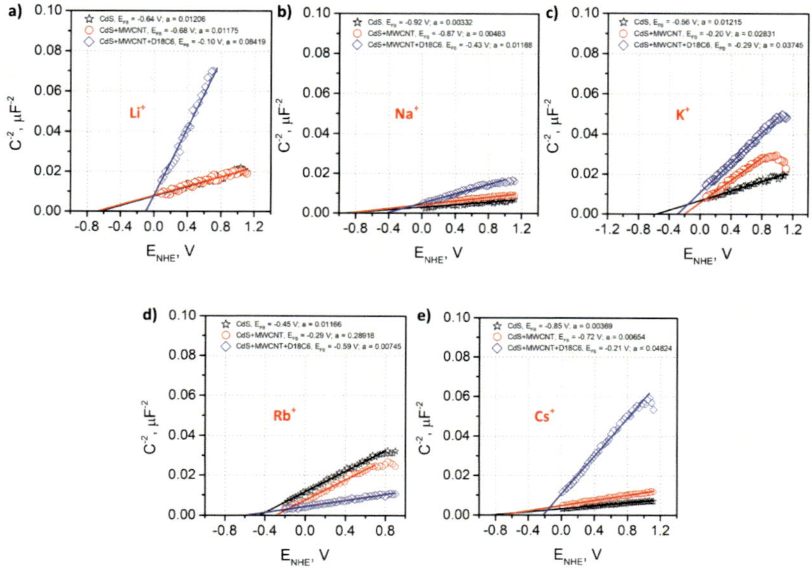

FIGURE 6

Ion type–dependent representation of M-S characteristics recorded for CdS-based electrodes (f=1kHz, purged with Ar (20 min), dark conditions).

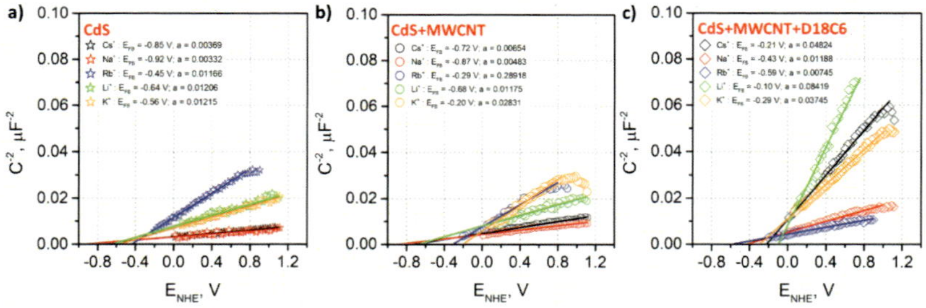

FIGURE 7

Materials composition – dependent representation of Mott-Schottky characteristics recorded in electrolytes containing different ions (f=1kHz, purged with Ar (20 min), dark conditions).

fication of CdS with MWCNT were also very negligible. Similar values and character of space charge layer capacitance changes resulted only in the slight changes in E_{FB} location. The slope of the Mott-Schottky plot indicated more noticeable differences in changes in space-charge capacitance with potential decreasing resulting from modification of CdS with MWCNT in the case of the electrolytes containing K^+, Rb^+, and Cs^+. Figure 7 summarises changes in Mott-Schottky characteristics resulting from the pres-

	CdS	CdS-MWCNT	CdS-MWCNT-D18C6
Li^+	-0.64	-0.68	-0.10
Na^+	-0.92	-0.87	-0.43
K^+	-0.56	-0.2	-0.29
Rb^+	-0.45	-0.29	-0.59
Cs^+	-0.85	-0.71	-0.21

TABLE 3
Values of E_{FB} (V) determined based on M-S plots obtained for several materials immersed in the electrolyte containing different ions (0.1 M).

ence of several ions in the space charge region and does not reveal any clear effect resulting from changes in the type of ions existing in the space charge region. The E_{FB} determined based on recorded Mott-Schottky plots are summarised in Table 3.

Both representations do not reveal any clear effect resulting from modification of the sensing material composition. There are no shifts of flat band location, Mott-Schottky slope, or space charge region capacitance, potential dependencies allowing utilisation of these materials toward sensing-related purposes. On the other hand, interactions of innocent metal cations with coordination compounds or nanomaterials usually follow a monotonous trend.[86-90] The character of changes presented in Figure 6 and Figure 7 indicates the complexity of the mechanisms that are behind the observed character of changes in E_{FB} and Mott-Schottky slope related to changes in electron donor concentration.

In the studied case, the lack of a monotonous trend indicates a complex interaction between three different semiconducting materials and simple aqueous electrolytes. Furthermore, attempts related to the combination of the obtained results with respect to the ionic radius (naked or hydrated) of the existing cations in the space charge region do not allow the estimation of clear E_{FB} shifts that would be useful in applications related to ion sensing (Figure 8a, b). Additionally, attempts to correlate E_{FB} with hydration energy or hydration number did not deliver satisfactory results (Figure 8c, d). DFT studies indicated that dibenzo-18-crown-6 exhibits the highest binding selectivity for K^+ ions compared to the remaining alkali metal cations.[82] The bond dissociation free energy (BDFE) was also reported to be strongly dependent on the solvation energy of M^+-D18C6, which may be crucial in the estimation of the binding selectivity of D18C6 in an aqueous solution. It was also pointed out that the non-electrostatic character dispersion interactions between solute and solvent are strongly dependent on the specific M^+-D18C6 structure and influence its solvation energies.[82]

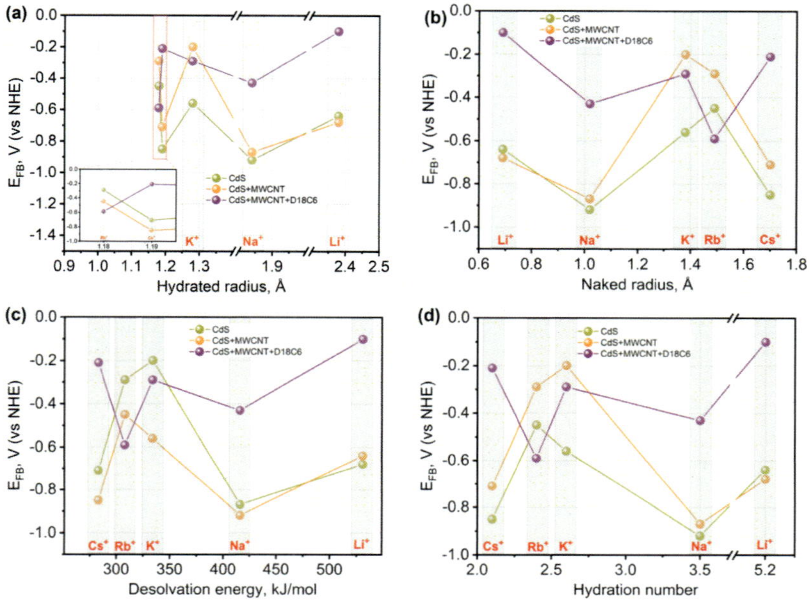

FIGURE 8
Dependence of the flat-band potential E_{FB} on a) ionic radius, b) the hydrated radius, c) desolvation energy, and d) the hydration number for synthesised sensing materials.

Although the above trend allows for a clear determination of the direction of changes in the E_{FB}, slope of the Mott-Schottky plot in the presence of individual ions (related to the size of the ion or its type) has not been observed. This fact may also be related to the non-linear change in the formation of solvation shells for individual alkali metal ions, which will have an indirect effect on the space-charge region capacity at the investigated material–electrolyte interface.[91-92]

The selection of appropriate, sometimes unconventional methods of data presentation may give new perspectives delivering valuable insights into the mechanisms responsible for observed effects. Figure 9 presents a visualisation of the results of flat-band potentials in a three-dimensional system. An attempt was made to check whether there is a nonmonotonic relationship between the presence of cations and the composition of the synthesised CdS-based materials. The E_{FB} values obtained from Mott-Schottky plots for three studied materials were treated as a set of points in three-dimensional space: (X, Y, Z), where each cation corresponds to one point. The relationship shows the differences in the E_{FB} localisation for all three materials. The 3D projection clearly shows that each sensor differentiates cations, but the explicit differentiation criterion is not known. Many factors influence the sensor response, among others, the solvation shells of the ions, the properties of the electrical double layer in different electrolytes,

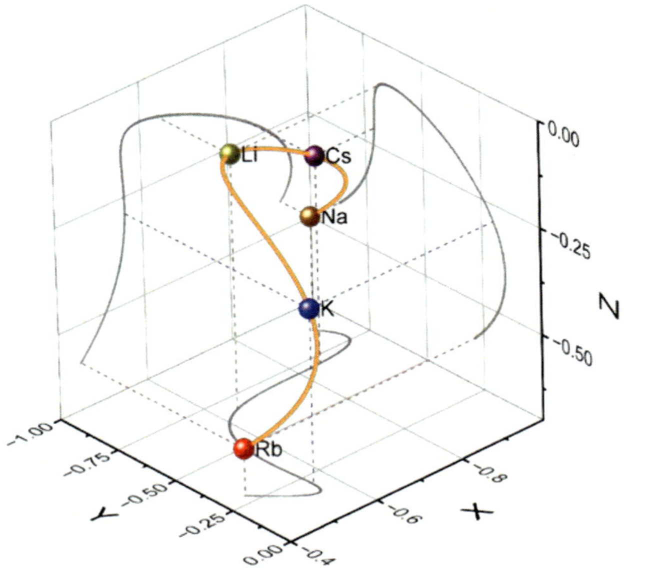

FIGURE 9
3D projection of the E_{FB} values for CdS-based materials immersed in an electrolyte containing various alkali metal ions (X, Y, Z – E_{FB} for CdS, CdS-MWCNT, CdS-MWCNT-D18C6, respectively).

and the affinity of ions of different radii for the sensing compounds. Moreover, the influence of other factors, such as the absorption capacity of particular types of ions on the CdS or MWCNT surface, may affect the sensing mechanism and thus EIS parameters. The scientific issue related to the absorption of alkali metal ions on the surface of cadmium sulphide or other *n*-type semiconductors has not been described yet. The obtained results are very interesting, however, only a precise determination of the factors responsible for changes in capacity in the space charge region may allow one to derive an equation describing the nature of changes in E_{FB} localisation – also in presented unconventional 3D space.

Interestingly, these experimental observations are not supported by simple density functional theory modelling, which does not differentiate various cations that participate in the formation of complexes with carbon nanotube-bound crown ethers (Figure 9). Density of states (DOS) calculations were performed for each of the tested systems to check whether the presence of the alkali metal ion affects the density of states distribution. In the DOS plots for three-component materials (CdS-MWCNT-D18C6) with the introduced alkali metal ion, one can notice a shift in the density of states towards lower energies (Figure 10 b-f) compared to the three-component material without the ion complex in the D18C6 structure (Figure 10a). Probably, after adding cations (especially the K^+ ion),

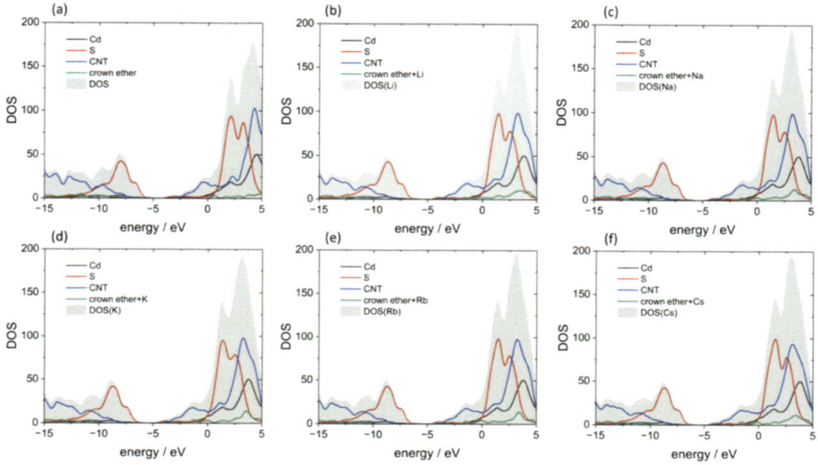

FIGURE 10
Total density of states (TDOS) distributions for the CdS/MWCNT/D18C6 ternary system without an ion in system a) and for the system with an alkali metal ion (b-f).

the energies of the unoccupied states shift to lower energies and become occupied. The interactions between the components of the system are electrostatic interactions; hence, there are no changes in the electronic structure (no new bonds are formed, and therefore no new electronic states appear).

We also illustrate the changes in charge distribution that occur after adding a cation to the CdS-MWCNT-D18C6 ternary system, which can be seen in Figure 11. Isosurface models of the system show the existence of clear π-π stacking of the nanotube and crown ether, as well as CdS molecules and carbon nanotube in each of the investigated cases (with or without a metal ion in the system). The charge is mostly located on the carbon nanotube, which, however, is mostly "electrophilic", that is, it has the ability to attract electrons, which is consistent with the literature.[93] Carbon nanotubes are materials that can produce π-π stacking bonds. Therefore, it was intuitive to expect the crown ether-MWCNT system to form this type of bond. However, it is interesting that in most cases adding a metal cation to the system does not change the potential distribution on the carbon nanotube, with the exception of adding the K^+ ion to the system, which is the ion that forms the most stable complex with D18C6 in an aqueous solution among all alkali metal ions. This is probably the result of a perfect fit into the crown ether gap by the solvated K^+ ion.[82,94] The formation of the D18C6-K^+ complex in aqueous solutions may change the charge distribution in the crown ether molecule itself and thus increase the electrophilic nature of the system (change the density of elec-

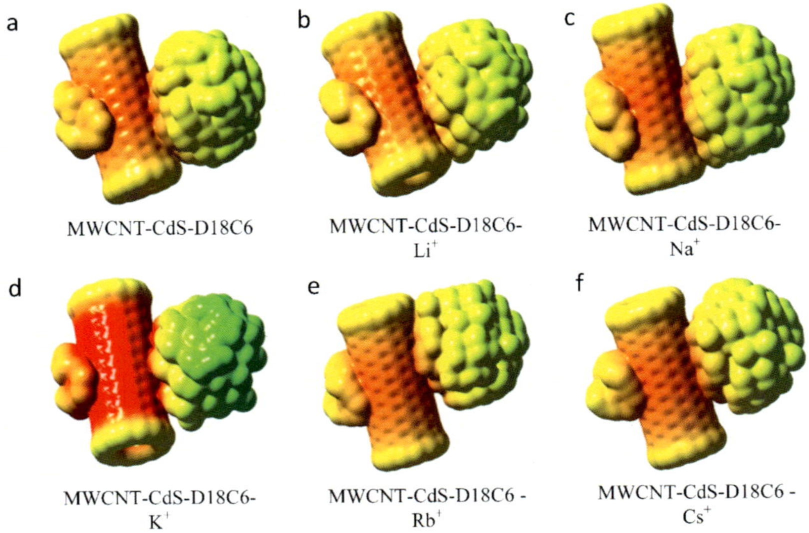

FIGURE 11
Electrostatic potential mapped onto the electron density isosurface with a colour map of the electrostatic potential distribution for a) a binary system (CdS-MWCNT) and a ternary system (CdS-MWCNT-D18C6), b) without added ion, c-g) with added Li^+, Na^+, K^+, Rb^+, and Cs^+ ion.

tronic states toward more positive ones). Theoretically, it could be concluded that the three-component system should be the most sensitive and selective for detecting potassium ions.

This ternary material was used to build a PEC sensor sensitive to the presence of alkaline ions in the solution. Moreover, they also wanted to check whether changing the system response would allow the identification of a particular ion. For this purpose, three-dimensional photocurrent maps were recorded for the obtained hybrid materials. The photocurrent map spectra for CdS-MWCNT-D18C6@ITO are presented in Figure 12. For each of the systems tested (CdS, CdS-MWCNT) and the sensor material (CdS-MWCNT-D18C6), the effect of switching photocurrents (PEPS effect) in oxygen-saturated solutions was observed. The values of the switching potentials for all materials are listed in Table 4.

For the electrode made from CdS, the PEPS effect occurs only at positive potentials (0 or 0.1 V), while for the two-component material (CdS-MWCNT) at zero potential (only in the presence of K^+ for +0.1 V). However, the situation changes when we measure the sensor material (CdS-MWCNT-D18C6). The results obtained show sensitivity to three metal cations: Na, K, and Rb, which is a consequence of the strongest interaction of these cations with the crown ether. In the presence of these ions, the PEPS effect occurs at a negative potential of - 0.1 V. It may be related to the shift of the density of states

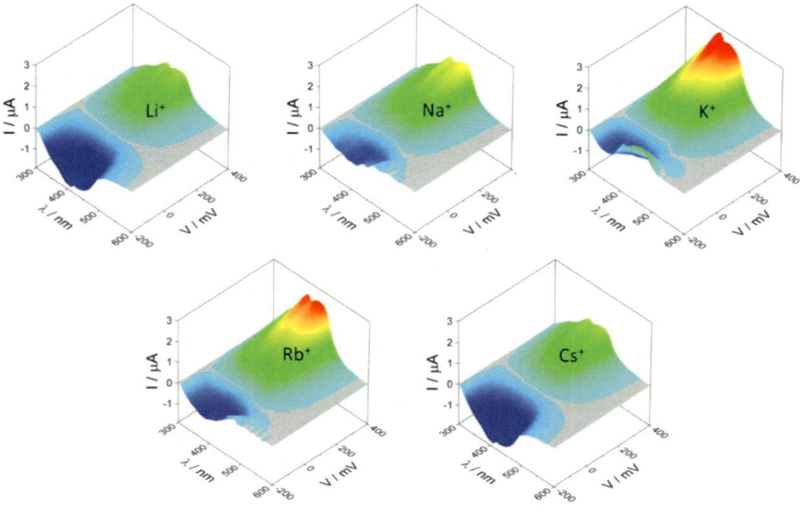

FIGURE 12
3D map of the photocurrent spectra registered for CdS/MWCNT/D18C6@ITO electrode in the presence of different cations.

Material	Switching potential / V				
	Li^+	Na^+	K^+	Rb^+	Cs^+
CdS	0.1	0.1	0.1	0.1	0
MWCNT-CdS	0	0	0.1	0	0
D18C6-MWCNT-CdS	0	-0.1	-0.1	-0.1	0

TABLE 4
Value of photocurrent switching potentials for each material in the presence of different ions.

toward lower energies in the presence of cations. Only in the case of Cs^+ ions, the PEPS effect occurs consistently at a potential of 0 V. This may indicate that caesium ions do not change the switching potential due to the lack/weak interactions of the ion with the surface of cadmium sulphide, carbon nanotubes, and D18C6. For Li^+ we do not observe the changes in switching potentials between two and three component materials (0 V). This may be due to its very strong solvation in aqueous solutions, so it has a large effective ionic radius that prevents the cation from "entering" the ether cavity. When comparing the values of anodic photocurrents, it can be seen that the highest photocurrent intensity was recorded for potassium and rubidium ions. The differences between them are not significant enough to clearly distinguish which cation we are dealing with.

6 CONCLUDING REMARKS

The examples discussed in this paper clearly indicate that there are numerous possibilities for novel sensing and computing approaches utilising cheap and ubiquitous materials. Everyday substances such as paper, milk, honey-derived materials, as well as cost-effective nanoparticles and nanotubes, can be successfully engaged in various information processing constructs: security devices, sensors, and neuromorphic devices. The variety of materials used in these devices as well as their functions and applications are limited only by the imagination and curiosity of researchers and engineers.

The integration of the described materials into electronic devices holds significant promise for addressing modern challenges in information technology. As our economy becomes increasingly dependent on digital solutions, the risk of digital exclusion, particularly among older generations, becomes a possibility. However, the development of low-cost, accessible, and sustainable electronics, such as biodegradable memristors and paper-based electronics, offers alternative solutions to this issue and promotes greater inclusivity in the digital world.

These innovative materials serve as examples of fully functional electronic devices that can be constructed even with limited resources, yet still supporting the non-von Neumann architecture for efficient information processing. While current commercialised technologies mostly rely on advanced fabrication techniques, the future lies in developing new hardware platforms that are sustainable and easy to fabricate.

ACKNOWLEDGEMENTS

The authors thank Dr. Marta Gajewska for transmission electron microscopy images. Dr. Kacper Pilarczyk for numerous discussions and critical comments. The authors acknowledge the financial support from the Polish National Science Centre within the OPUS programme (grant agreement No. UMO-2020/37/B/ST5/00728). This research was partly supported by the programme 'Excellence initiative-research university' for the AGH University of Science and Technology. The authors gratefully acknowledge Polish high-performance computing infrastructure PLGrid (HPC Centres: ACK Cyfronet AGH) for providing computer facilities and support within computational grant no. PLG/2024/017004.

REFERENCES

[1] Tomczyńska, W., Digital exclusion: definitions, causes, countermeasures. *Adeptus* 2017, 10, 1503.

[2] Juszczyk, S.; Kim, S., Digital Exclusion in Poland and South Korea - A Comparative Analysis of the Phenomenon. *New Educ. Rev.* 2022, 68, 32-47.

[3] https://companiesmarketcap.com/. (accessed July 5th, 2024).

[4] Andrae, A. S. G.; Edler, T., On Global Electricity Usage of Communication Technology: Trends to 2030. *Challenges* 2015, 6 (1), 117-157.

[5] Jones, N., The information factories. *Nature (London)* 2018, 561, 163-166.

[6] Binnemans, K.; Jones, P. T.; Müller, T.; Yurramendi, L., Rare Earths and the Balance Problem: How to Deal with Changing Markets? *Journal of Sustainable Metallurgy* 2018, 4 (1), 126-146.

[7] Balaram, V., Rare earth elements: A review of applications, occurrence, exploration, analysis, recycling, and environmental impact. *Geoscience Frontiers* 2019, 10 (4), 1285-1303.

[8] Kirkpatrick, K., Electronics need rare earths. *Commun. ACM* 2019, 62 (3), 17–18.

[9] Wolpert, D. H.; Macready, W. G., No free lunch theorems for optimization. *IEEE Transactions on Evolutionary Computation* 1997, 1 (1), 67-82.

[10] Lis, M.; Onuma, S.; Przyczyna, D.; Zawal, P.; Mazur, T.; Pilarczyk, K.; Gentili, P. L.; Kasai, S.; Szaciłowski, K., From Oscillatory Reactions to Robotics: A Serendipitous Journey Through Chemistry, Physics and Computation. In *Handbook of Unconventional Computing*, World Scientific: 2021; pp 1-79.

[11] Vopson, M. M., The information catastrophe. *AIP Advances* 2020, 10 (8), 085014.

[12] Lem, S., *The Star Diaries*. Pengin Books: London, 2021.

[13] Wlaźlak, E.; Przyczyna, D.; Gutierrez, R.; Cuniberti, G.; Szaciłowski, K., Towards synthetic neural networks: can artificial electrochemical neurons be coupled with artificial memristive synapses? *Jpn. J. Appl. Phys.* 2020, 59 (SI), SI0801.

[14] Schuster, A. J., *Understanding information: from the Big Bang to Big Data*. Springer: Cham, 2017.

[15] Zgurovsky, M. Z.; Zaychenko, Y. P., *Big Data: Conceptual Analysis nad applications*. Springer Nature: Cham, Switzerland, 2020.

[16] López, C., Artificial Intelligence and Advanced Materials. *Adv. Mater.* 2023, 35 (23), 2208683.

[17] Lloyd, S., Computational capacity of the Universe. *Phys. Rev. Lett.* 2002, 88, 237901.

[18] Moore, G. E., Cramming more components onto integrated circuits. *Electronics* 1965, 38, 114.

[19] Dennard, R. H.; Gaensslen, F. H.; Yu, H. N.; Rideout, V. L.; Bassous, E.; LeBlanc, A. R., Design of ion-implanted MOSFET's with very small physical dimensions. *IEEE J. Solid-State Circ.* 1974, 9 (5), 256-268.

[20] Amdahl, G. M., Validity of the Single Processor Approach to Achieving Large Scale Computing Capabilities, Reprinted from the AFIPS Conference Proceedings, Vol. 30 (Atlantic City, N.J., Apr. 18–20), AFIPS Press, Reston, Va., 1967, pp. 483–485, when Dr. Amdahl was at International Business Machines Corporation, Sunnyvale, California. *IEEE Solid-State Circuits Society Newsletter* 2007, 12 (3), 19-20.

[21] Gustafson, J. L., Reevaluating Amdahl's law. *Commun. ACM* 1988, 31 (5), 532–533.

[22] Kimovski, D.; Saurabh, N.; Jansen, M.; Aral, A.; Al-Dulaimy, A.; Bondi, A. B.; Galletta, A.; Papadopoulos, A. V.; Iosup, A.; Prodan, R., Beyond Von Neumann in the Computing Continuum: Architectures, Applications, and Future Directions. *IEEE Internet Computing* 2024, 28 (3), 6-16.

[23] Arikpo, I.; Ogban, F.; Eteng, I., Von neumann architecture and modern computers. *Global Journal of Mathematical Sciences* 2007, 6 (2), 97-103.

[24] Jouppi, N.; Young, C.; Patil, N.; Patterson, D., A domain-specific architecture for deep neural networks. *Communications of the ACM* 2018, 61, 50-59.

[25] Rupp, K. 42 Years of Microprocessor Trend Data. https://www.karlrupp.net/2018/02/42-years-of-microprocessor-trend-data/ (accessed 18.07.2022).

[26] Landauer, R., Irreversibility and Heat Generation in the Computing Process. *IBM J. Res. Dev.* 1961, 5, 183-191.

[27] Einstein, A., *Über die spezielle und die allgemeine Relativitätstheorie*. Friedrich Vieweg & Sohn: Braunschweig, 1919.

[28] Szaciłowski, K., *Infochemistry. Information Processing at the Nanoscale*. John Wiley & Sons: Chichester, 2012.

[29] Vopson, M. M., The mass-energy-information equivalence principle. *AIP Advances* 2019, 9 (9), 095206.

[30] Bormashenko, E., The Landauer Principle: Re-Formulation of the Second Thermodynamics Law or a Step to Great Unification? *Entropy* 2019, 21 (10), 918.

[31] Herrera, L., The mass of a bit of information and the Brillouin's principle. *Fluctuation and Noise Letters* 2014, 13 (01), 1450002.

[32] Burgin, M.; Mikkilineni, R., Is Information Physical and Does It Have Mass? *Information* 2022, 13 (11), 540.

[33] Aguirre, A.; Foster, B.; Merali, Z., *It From Bit or Bit From It? On Physics and Information*. Springer International Publishing: Cham, 2015.

[34] Wheeler, J. A., Information, physics, quantum: Ther search for links. *Proc. 3rd Int. Symp. Foundations of Quantum Mechanic* 1989, 354-386.

[35] de Silva, A. P., *Molecular Logic-based Computation*. The Royal Society of Chemistry: Cambridge, 2012.

[36] Katz, E., *Molecular and supramolecular information processing*. Wiley-VCH: Weinheim, 2012.

[37] Katz, E., *Biomolecular information processing*. Wiley-VCH: Weinheim, 2012.

[38] Rehman, M. M.; ur Rehman, H. M. M.; Kim, W. Y.; Sherazi, S. S. H.; Rao, M. W.; Khan, M.; Muhammad, Z., Biomaterial-Based Nonvolatile Resistive Memory Devices toward Ecofriendliness and Biocompatibility. *ACS Applied Electronic Materials* 2021, 3 (7), 2832-2861.

[39] Chen, Y.-C.; Yu, H.-C.; Huang, C.-Y.; Chung, W.-L.; Wu, S.-L.; Su, Y.-K., Nonvolatile Bio-Memristor Fabricated with Egg Albumen Film. *Sci. Rep.* 2015, 5 (1), 10022.

[40] Zhou, G.; Sun, B.; Zhou, A.; Wu, B.; Huang, H., A larger nonvolatile bipolar resistive switching memory behaviour fabricated using eggshells. *Current Applied Physics* 2017, 17 (2), 235-239.

[41] He, X.; Zhang, J.; Wang, W.; Xuan, W.; Wang, X.; Zhang, Q.; Smith, C. G.; Luo, J., Transient Resistive Switching Devices Made from Egg Albumen Dielectrics and Dissolvable Electrodes. *ACS Applied Materials & Interfaces* 2016, 8 (17), 10954-10960.

[42] Zhu, J. X.; Zhou, W. L.; Wang, Z. Q.; Xu, H. Y.; Lin, Y.; Liu, W. Z.; Ma, J. G.; Liu, Y. C., Flexible, transferable and conformal egg albumen based resistive switching memory devices. *RSC Advances* 2017, 7 (51), 32114-32119.

[43] Nagashima, K.; Koga, H.; Celano, U.; Zhuge, F.; Kanai, M.; Rahong, S.; Meng, G.; He, Y.; De Boeck, J.; Jurczak, M.; Vandervorst, W.; Kitaoka, T.; Nogi, M.; Yanagida, T., Cellulose Nanofiber Paper as an Ultra Flexible Nonvolatile Memory. *Scientific Reports* 2014, 4 (1), 5532.

[44] Hosseini, N. R.; Lee, J.-S., Biocompatible and Flexible Chitosan-Based Resistive Switching Memory with Magnesium Electrodes. *Advanced Functional Materials* 2015, 25 (35), 5586-5592.

[45] Hung, Y.-C.; Hsu, W.-T.; Lin, T.-Y.; Fruk, L., Photoinduced write-once read-many-times memory device based on DNA biopolymer nanocomposite. *Applied Physics Letters* 2011, 99 (25), 253301.

[46] Sueoka, B.; Hasan Tanim, M. M.; Williams, L.; Xiao, Z.; Seah, Y. Z.; Cheong, K. Y.; Zhao, F., A synaptic memristor based on natural organic honey with neural facilitation. *Organic Electronics* 2022, 109, 106622.

[47] Kim, J.; Dowling, V. J.; Datta, T.; Pershin, Y. V., Whisky-Born Memristor. *Physica Status Solidi (a)* 2023, 220 (11), 2200643.

[48] Sharma, N.; Singh, K.; Tripathi, C. C.; Bera, M. K., Biomemristor based on a natural medicinal plant (Tinospora cordifolia) and their phototunable resistive switching properties integrated with carbon quantum dots. *Applied Physics A* 2022, 129 (1), 58.

[49] Schneider, E. M.; Bärtsch, A.; Stark, W. J.; Grass, R. N., Safe One-Pot Synthesis of Fluorescent Carbon Quantum Dots from Lemon Juice for a Hands-On Experience of Nanotechnology. *Journal of Chemical Education* 2019, 96 (3), 540-545.

[50] Khan, S. M.; Nassar, J. M.; Hussain, M. M., Paper as a Substrate and an Active Material in Paper Electronics. *ACS Applied Electronic Materials* 2021, 3 (1), 30-52.

[51] Conceição, J. H. F.; Party, M.; Curtil, D.; Švecová, L.; Marlin, N.; Reverdy-Bruas, N., Investigation of methods to quantify silver screen-printed onto cellulosic substrate:

towards recycling of printed electronics. *Flexible and Printed Electronics* 2023, 8 (3), 035014.

[52] Wang, Z.; Tammela, P.; Huo, J.; Zhang, P.; Strømme, M.; Nyholm, L., Solution-processed poly(3,4-ethylenedioxythiophene) nanocomposite paper electrodes for high-capacitance flexible supercapacitors. *Journal of Materials Chemistry A* 2016, 4 (5), 1714-1722.

[53] Morais, R.; Vieira, D. H.; Gaspar, C.; Pereira, L.; Martins, R.; Alves, E. N., Influence of paper surface characteristics on fully inkjet printed PEDOT:PSS-based electrochemical transistors. *Semiconductor Science and Technology* 2021, 36 (12), 125005.

[54] Liu, K.; Ouyang, B.; Guo, X.; Guo, Y.; Liu, Y., Advances in flexible organic field-effect transistors and their applications for flexible electronics. *NPJ Flexible Electronics* 2022, 6 (1), 1.

[55] Lee, S.; Hong, Y.; Shim, B. S., Biodegradable PEDOT:PSS/Clay Composites for Multi-functional Green-Electronic Materials. *Advanced Sustainable Systems* 2022, 6 (2), 2100056.

[56] Bagdžiūnas, G.; Palinauskas, D., Poly(9H-carbazole) as a Organic Semiconductor for Enzymatic and Non-Enzymatic Glucose Sensors. *Biosensors* 2020, 10 (9), 104.

[57] Tong, C., Substrate and Encapsulation Materials for Printed Flexible Electronics. In *Advanced Materials for Printed Flexible Electronics*, Tong, C., Ed. Springer International Publishing: Cham, 2022; pp 221-255.

[58] Choi, W.; Shin, J.; Kim, Y. J.; Hur, J.; Jang, B. C.; Yoo, H., Versatile Papertronics: Photo-Induced Synapse and Security Applications on Papers. *Advanced Materials* 2024, 36 (38), 2312831..

[59] Gao, Y.; Choi, S., Stepping Toward Self-Powered Papertronics: Integrating Biobatteries into a Single Sheet of Paper. *Advanced Materials Technologies* 2017, 2 (1), 1600194..

[60] Hasan, M. R.; Sharma, P.; Suleman, S.; Mukherjee, S.; Celik, E. G.; Timur, S.; Pilloton, R.; Narang, J., Papertronics: Marriage between Paper and Electronics Becoming a Real Scenario in Resource-Limited Settings. *Acs Applied Bio Materials* 2023, 6 (4), 1368-1379.

[61] Yalagala, B.; Sahatiya, P.; Mattela, V.; Badhulika, S., Ultra-low Cost, Large Area Graphene/MoS₂-Based Piezotronic Memristor on Paper: A Systematic Study for Both Direct Current and Alternating Current Inputs. *ACS Applied Electronic Materials* 2019, 1 (6), 883-891.

[62] Wiklund, J.; Karakoç, A.; Palko, T.; Yiğitler, H.; Ruttik, K.; Jäntti, R.; Paltakari, J., A Review on Printed Electronics: Fabrication Methods, Inks, Substrates, Applications and Environmental Impacts. 2021, 5 (3), 89.

[63] Skarżyński, K.; Krzemiński, J.; Jakubowska, M.; Słoma, M., Highly conductive electronics circuits from aerosol jet printed silver inks. *Scientific Reports* 2021, 11 (1), 18141.

[64] Anushka; Bandopadhyay, A.; Das, P. K., Paper based microfluidic devices: a review of fabrication techniques and applications. *The European Physical Journal Special Topics* 2023, 232 (6), 781-815.

[65] Yalagala, B. P.; Sahatiya, P.; Kolli, C. S. R.; Khandelwal, S.; Mattela, V.; Badhulika, S., V_2O_5 Nanosheets for Flexible Memristors and Broadband Photodetectors. *ACS Applied Nano Materials* 2019, 2 (2), 937-947.

[66] Enaganti, P. K.; Selamneni, V.; Sahatiya, P.; Goel, S., MoS2/cellulose paper coupled with SnS2 quantum dots as 2D/0D electrode for high-performance flexible supercapacitor. *New Journal of Chemistry* 2021, 45 (19), 8516-8526.

[67] Veeralingam, S.; Sahatiya, P.; Badhulika, S., Papertronics: Hand-Written MoS on Paper Based Highly Sensitive and Recoverable Pressure and Strain Sensors. *IEEE Sensors Journal* 2021, 21 (7), 8943-8949.

[68] Korotcenkov, G.; Simonenko, N. P.; Simonenko, E. P.; Sysoev, V. V.; Brinzari, V., Paper-Based Humidity Sensors as Promising Flexible Devices, State of the Art, Part 2: Humidity-Sensor Performances. *Nanomaterials* 2023, 13 (8), 1381.

[69] Mohammadifar, M.; Choi, S., A Papertronic, On-Demand and Disposable Biobattery: Saliva-Activated Electricity Generation from Lyophilized Exoelectrogens Preinoculated on Paper. *Advanced Materials Technologies* 2017, 2 (9), 1700127.

[70] Yetisen, A. K.; Akram, M. S.; Lowe, C. R., Paper-based microfluidic point-of-care diagnostic devices. *Lab on a Chip* 2013, 13 (12), 2210-2251.

[71] Martinez, A. W.; Phillips, S. T.; Whitesides, G. M.; Carrilho, E., Diagnostics for the Developing World: Microfluidic Paper-Based Analytical Devices. *Analytical Chemistry* 2010, 82 (1), 3-10.

[72] Pungjunun, K.; Nantaphol, S.; Praphairaksit, N.; Siangproh, W.; Chaiyo, S.; Chailapakul, O., Enhanced sensitivity and separation for simultaneous determination of tin and lead using paper-based sensors combined with a portable potentiostat. *Sensors and Actuators B: Chemical* 2020, 318, 128241.

[73] Fang, Z.; Zhu, H.; Yuan, Y.; Ha, D.; Zhu, S.; Preston, C.; Chen, Q.; Li, Y.; Han, X.; Lee, S.; Chen, G.; Li, T.; Munday, J.; Huang, J.; Hu, L., Novel Nanostructured Paper with Ultrahigh Transparency and Ultrahigh Haze for Solar Cells. *Nano Letters* 2014, 14 (2), 765-773.

[74] Xiong, C.; Wang, T.; Han, J.; Zhang, Z.; Ni, Y., Recent Research Progress of Paper-Based Supercapacitors Based on Cellulose. *Energy & Environmental Materials* 2024, 7 (3), e12651.

[75] Jang, B. C.; Kim, S.; Yang, S. Y.; Park, J.; Cha, J.-H.; Oh, J.; Choi, J.; Im, S. G.; Dravid, V. P.; Choi, S.-Y., Polymer Analog Memristive Synapse with Atomic-Scale Conductive Filament for Flexible Neuromorphic Computing System. *Nano Letters* 2019, 19 (2), 839-849.

[76] Pushparaj, V. L.; Shaijumon, M. M.; Kumar, A.; Murugesan, S.; Ci, L.; Vajtai, R.; Linhardt, R. J.; Nalamasu, O.; Ajayan, P. M., Flexible energy storage devices based on nanocomposite paper. *Proceedings of the National Academy of Sciences* 2007, 104 (34), 13574-13577.

[77] Landers, M.; Elhadad, A.; Rezaie, M.; Choi, S., Integrated Papertronic Techniques: Highly Customizable Resistor, Supercapacitor, and Transistor Circuitry on a Single Sheet of Paper. *ACS Applied Materials & Interfaces* 2022, 14 (40), 45658-45668.

[78] Selamneni, V.; Kanungo, S.; Sahatiya, P., Large area growth of SnS2/graphene on cellulose paper as a flexible broadband photodetector and investigating its band structure through first principles calculations. *Materials Advances* 2021, 2 (7), 2373-2381.

[79] Konkoli, Z., Reservoir Computing. In *Encyclopedia of Complexity and Systems Science*, Meyers, R. A., Ed. Springer Berlin Heidelberg: Berlin, Heidelberg, 2017; pp 619-629.

[80] Adamatzky, A.; Szaciłowski, K.; Konkoli, Z.; Werner, L. C.; Przyczyna, D.; Sirakoulis, G. C., On buildings that compute. A proposal. In *From astrophysics to unconventional computation*, Adamatzky, A.; Kendon, V., Eds. Springer Nature Switzerland AG: Cham, Switzerland, 2020.

[81] Przyczyna, D.; Suchecki, M.; Adamatzky, A.; Szaciłowski, K., Towards Embedded Computation with Building Materials. *Materials* 2021, 14 (7) 1724.

[82] Choi, C. M.; Heo, J.; Kim, N. J., Binding selectivity of dibenzo-18-crown-6 for alkali metal cations in aqueous solution: A density functional theory study using a continuum solvation model. *Chemistry Central Journal* 2012, 6 (1), 84.

[83] Sahu, P.; Ali, S. M.; Singh, J. K., Structural and dynamical properties of Li+-dibenzo-18-crown-6(DB18C6) complex in pure solvents and at the aqueous-organic interface. *J. Mol. Model.* 2014, 20 (9), 2413.

[84] Grootenhuis, P. D. J.; Kollman, P. A., Molecular mechanics and dynamics studies of crown ether - cation interactions: free energy calculations on the cation selectivity of dibenzo-18-crown-6 and dibenzo-30-crown-10. *J. Am. Chem. Soc.* 1989, 111 (6), 2152-2158.

[85] Pranowo, H. D.; Anwar, C., Interaction between Li+ cation with crown ethers of bz15c5, dbz16c5 and dbz18c6: molecular modeling base on MNDO/d semiempirical method. *Ind. J. Chem.* 2003, 3, 55-66.

[86] Szaciłowski, K.; Stochel, G.; Stasicka, Z.; Kisch, H., Reaction of sodium pentacyanonitrosylferrate(2−) with aliphatic thiolates. Synthesis and properties of $[Fe(CN)_5N(O)SR]^{3-}$ products. *New. J. Chem.* 1997, 21, 893-902.

[87] Szaciłowski, K.; Macyk, W.; Stochel, G.; Stasicka, Z.; Sostero, S.; Traverso, O., Ligand and medium controlled photochemistry of iron and ruthenium mixed-ligand complexes: prospecting for versatile systems. *Coord. Chem. Rev* . 2000, 208, 277-297.

[88] Lawrance, G. A. M., M.; Neuhold, Y.-M.; Szaciłowski, K.; Barbieri, A.; Stasicka, Z., An electrospray ionization mass spectrometry study of the nitroprusside–cation–thiolate system *J. Chem. Soc., Dalton Trans.* 2002, 3649-3655.

[89] Szaciłowski, K., Molecular Logic Gates Based on Pentacyanoferrate Complexes: From Simple Gates to Three-Dimensional Logic Systems. *Chem. Eur. J.* 2004, 10, 2520-2528.

[90] Suchecki, M.; Klejna, S.; Marciszko-Wiąckowska, M.; Bednarski, W.; Rosiak, N.; Lewandowska, K.; Szaciłowski, K., Supercapacitance in graphene oxide materials modified with tetrapyrrole dyes: a mechanistic study. *Nanoscale* 2022, 14 (23), 8534-8547.

[91] Hou, Z.; Dong, M.; Xiong, Y.; Zhang, X.; Zhu, Y.; Qian, Y., Formation of Solid–Electrolyte Interfaces in Aqueous Electrolytes by Altering Cation-Solvation Shell Structure. *Advanced Energy Materials* 2020, 10 (15), 1903665.

[92] Mähler, J.; Persson, I., A Study of the Hydration of the Alkali Metal Ions in Aqueous Solution. *Inorganic Chemistry* 2012, 51 (1), 425-438.

[93] Zhang, Y.; Wang, D.; Wang, W., Beyond the σ-hole and π-hole: The origin of the very large electrophilic regions of fullerenes and carbon nanotubes. *Computational and Theoretical Chemistry* 2018, 1128, 56-59.

[94] Nasri, F., A Review of the Thermodynamics of Complexation of Crown Ethers With Metal Ion. *Journal of Advances in Environmental Health Research* 2022, 10 (4), 263-272.

How to Build Controllable Biomolecular Circuits Using Protein Filaments

JACK A. TUSZYNSKI

Department of Physics, University of Alberta, Edmonton, Alberta, Canada
and
DIMEAS, Politecnico di Torino, Turin, Italy

Received: September 1, 2024. Accepted: September 28, 2024.

The structure-function relationship is the basis of quantitative analysis of living organisms whose fundamental unit is a cell. Cellular structural and functional complexity is a challenge to our understanding of responses to various environmental signals and stimuli. Electrical and electromagnetic interactions within and between cells are particularly poorly understood. Experimental investigations in parallel with computational modelling using modern methods and tools are gradually developing towards an integrated model of the cell as bioelectric circuitry. A complete quantitative bioelectric model of various subcellular components and a whole cell could allow us to reverse engineer the underlying bio-electrodynamic design principles and use this information for the construction of novel bioelectric devices. As a result, controllable use of such devices for the development of hybrid technologies and within biological systems can allow us to manipulate cell differentiation, cell division and other processes. While much is known about the electric properties of cell membranes, explorations of the cytoskeleton are still nebulous. Key cytoskeleton components, actin filaments and microtubules, play essential roles in cell motility, mitosis, cell differentiation, transport and signaling. Their elementary protein building blocks self-assemble into cell-spanning filaments, and are strongly affected by temperature, ionic concentrations, pH and other factors such as pharmacological agents. These factors, affecting cellular structure formation, also affect cellular responses to electric and electromagnetic fields in a largely unexplored way. Future research in this area of investigations may have major implications for the development of novel therapeutic modalities and for a range of nanotechnology applications such as nano-sensors and biocomputing elements.

Keywords: Tubulin, microtubules, electrical circuits, cytoskeleton

*Corresponding author email: jack.tuszynski@gmail.com

1 INTRODUCTION

Research on nano-devices is aimed at learning the physical properties of nanometer-scale structures and then determining how these can be used to perform various functions such as computing. Current nanocomputing research includes the study of very small electronic devices and biomolecules, their fabrication, architectures and physical characterization. Nanostructures studied so far include semiconductor quantum dots, single electron structures, nanomagnetic logic devices, and various organic molecules. For a successful development of a new technology based on such structures, a number of significant barriers must be overcome including precise control of their architecture, physical functions and scalability. Furthermore, the main obstacles to progress can be classified into: (a) hardware challenges, i.e., the physical composition of a nano-computer, its architecture, its communications structure, and the associated peripherals, and (b) software challenges, i.e. creating new software, operating systems, and utilities that must be designed and developed, enabling nano-scale computer chips to be integrated into computational devices in order to execute the desired operations in a given environment. Much of the progress in this area can be achieved by using biological materials or hybrid structures combining biological and inorganic components. This paper is intended to provide a small step in the direction of developing controllable logical elements based on cellular components.

The fundamental unit of living organisms is a cell whose daunting structural and dynamical complexity makes our understanding of its responses to environmental changes a major challenge to modern science. Biology alone cannot address in quantitative and predictive terms the molecular mechanisms underlying these responses, especially those pertaining to electrical and electromagnetic (EM) interactions with living cells. Through careful experiments and parallel computational modeling major insights into the functioning of living cells can be made when viewed through the lens of complex bioelectric circuit systems, i.e. using the methods of nanoscience, biophysics and bioengineering. The development of an integrated model of a cell as a bioelectric circuit can provide a blueprint for physical interventions into the state of the cell. We first need to carefully investigate the bio-electric properties of the cell's major components as independent entities and then the cell as a whole. This may allow the reverse engineering of the bioelectric and bio-electrodynamic design principles to be able to control the state of the cell to advance both basic science as well as medical and nanotechnological applications as a consequence. The latter would allow for control of such processes as cell differentiation and cell division with obvious implications for therapeutic purposes, e.g. cancer treatment. Scaling it up to cell and tissue culture can lead to generation of artificial organs and prosthetic devices.

Additionally, a radically new science-enabled technology for the production of adaptive learning self-growing nano-computers can be achieved using protein and hybrid polymer chips. These polymer chips can be made of biological polymers which support ionic conductivity, namely actin filaments (AFs), microtubules (MTs), intermediate filaments (IFs) and their associated cross-linking proteins (microtubules associated proteins- MAPs [1] and actin related proteins – ARPs) with the addition of DNA, RNA and synthetic polymers. These structures exhibit non-trivial electronic, solitonic, excitonic and ionic conduction properties. Such polymer networks can have optical and electrical I/O elements and can process information via interactions involving, for example, travelling ionic condensations, dynamical changes of conductivity and electro-mechanical couplings [2-3]. It is hoped that this can be implemented in a highly efficient manner as a high-performance massively-parallel nano/microscale computer. Architectures of polymer computing chips can be grown via chemically, hydrodynamically, electrically or electromagnetically (e.g., optically) controllable polymerisation reactions. These reactions may also include additional proteins such as MAPs or ARPs to produce stable protein filament networks with complex geometries, for example using electrical stimulation and/or by imposing geometrical constraints on the physical space using patterned solid surfaces. Fig. 1 illustrates the already known tubulin-based macromolecular polymeric cellular structures that can be useful in this regard.

Furthermore, state-of-art achievements of nanotechnology, molecular biology, unconventional computing, polymer chemistry, novel hardware, advanced functional materials, electrophysiology, theory of computation, and electronic engineering can be used to develop computer models and experi-

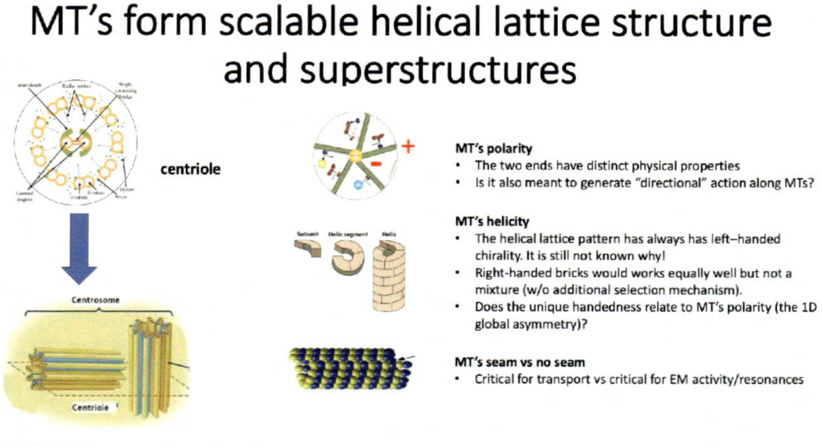

FIGURE 1
An illustration of tubulin-based sub-cellular structures and their geometries.

mental laboratory prototypes of self-growing computing devices built from polymers networks. The following characteristics are envisaged as desirable for polymer computers: (1) computation can be implemented via dynamical reprogramming of structural and electrical properties of polymer networks; (2) polymers can be self-growing via controlled polymerisation programming by patterning surfaces with DNA, binding proteins (e.g. biotin and streptavidin), electrical fields and geometrical constraints; (3) polymer fibres are able to integrate massive amounts of complex spatial and time-varying stimuli to effect local changes in electrical charge distribution, mechanical conformations and space-time dynamics of polymerisation; (4) the I/O can be implemented via electrical, chemical, mechanical and optical means. This can lead to ground-breaking designs, implementations and prototypes of cytoskeleton-based computers and computer chips. There is a potential for major advances in computer science (algorithms/architectures of molecular computing), computational geometry (polymer-based morphological processors), general purpose nano-scale computing, biology (utilizing the properties of biopolymers at the interface with electronics), material science (functional elements), electronics (high density nano-circuits), self-assembly regenerative systems, and last but not least, molecular neuromorphic devices.

This can also lead to the uncovering of a yet-unknown electrodynamic design blueprint for eukaryotic cells, both generically and in a comparative way between normal and cancer cells. Indeed, cancer cells exhibit a large number of major changes in their electro-chemical properties when compared to normal cells (e.g. pH, transmembrane potential, ionic concentrations, etc.) which are yet to be exploited therapeutically. For this to happen, we need to understand these properties in a quantitative manner, which will enable the development of novel methods of manipulating cells using electric and EM fields, for example transforming stem cells into specific phenotypes. Pathological states of cells (e.g. malignant) coud then be either forced into senescence, apoptosis or, preferably, back to their normal phenotype. While no clear mechanism of action is known yet, clinical applications of tumor-treating fields (TTFields), i.e. ~100 kHz electric fields [4-6] offer an enormous promise in the fight against cancer. Another example is the use of transmembrane potential modulation via specific chemicals to regulate cell differentiation. However, we need to systematically investigate a range of electric and EM field effects on cells and subcellular structures as functions of field intensity, modulation frequency, time duration of the applied external fields and ambient conditions of pH [7], ionic concentrations and temperature. These external stimuli are known to affect the response of various biological systems but not in a quantitatively predictive detail. Importantly, these responses also vary between cell types, which requires careful investigations. To understand how cells respond to external fields, we must delve into cell interiors, especially their cytoskeleton structures. Particularly important cytoskeleton proteins: tubulin and their complexes, have unique biophysical and biochemical properties, such as very large electric charges and dipole moments both permanent and induced [8-10]. Therefore, they are the primary intracellular targets for strong interactions with electric and EM fields.

Photobiological factors involve a specific type of electromagnetic field (i.e. in the range from the lower part of the UV spectrum (approx. 10 nm) to the upper part of the IR (approx. 1 mm). Moreover, interconnections between the subcellular elements such as MTs, AFs and MAPs are of potential use as controllable computational units in complex circuits. EM fields of the cells can affect both function and structure, hence should play an important role in the process of self-organization. One of the manifestations of the non-equilibrium behavior of living cells is the permanent production of electron excitation in biomolecules leading to ultra-weak photon emission. Biological ultra-weak photon emission is detectable from every metabolically active biological species under suitable conditions. The electrodynamic activity of living cells involves a broad range of frequencies, from kHz to THz. to infrared, visible and even ultraviolet ranges. These frequency ranges are related to electromechanical vibrations of subcellular structures. It is hypothesized that EM fields generated by sub-cellular coupled oscillations, for example of MTs, contribute significantly to biological self-organization and both intra- and inter-cellular communication [12-13].

MTs have unique biophysical and biochemical properties, such as very large electric dipole moments both permanent and induced. Theoretical investigations of these properties are supported by the growing body of experimental evidence on the different ways environmental perturbations (pH, ionic concentrations and gradients, electric and electromagnetic fields, mechanical stimuli) affect MTs and their mutual interaction at a distance. While diffusion facilitates short-range interactions and stochastic encounters between cellular components, the formation of complexes from protein constituents must involve mechanisms that are not exclusively diffusion-driven. Short-range specific interactions importantly determine the behavior of biological systems *via* the so-called "lock and key" principle of shape matching between the cognate molecules [14]. However, there must also be long-range effects [15], which are sensitive to electric and EM fields. MTs and MT constructs (centrioles, MT bundles in neurons, cilia, flagella, MT-mitochondria complexes, etc.) are critical to the function of all eukaryotic cells, i.e., cell division, intracellular transport, ion channel regulation, cell motility, and signaling. Recently, their electrical conduction properties indicate their role as potential EM energy transducers. The structure of MT-based constructs indicates their potential role in specialized function including cell-cell communication. This knowledge will also lead to precisely-designed nano-bio-technology devices and their complex architectures for numerous applications including sensors, self-replicating, self-repairing bioelectronic devices, biocomputing elements, etc.

2 BIOPHYSICAL PROPERTIES OF THE COMPONENT UNITS

While much is known about the electric properties of cellular membranes (with ion channels) representing a leaky capacitor, explorations of the corresponding properties of the key subcellular components, especially the cytoskeleton, are still nebulous. Importantly, the organization of cellular structures in eukaryotic cells is primarily governed by cytoskeleton proteins [14]. In particular, actin filaments (AFs) and microtubules (MTs) play essential roles in cell motility, cytokinesis, mitosis, cell differentiation, intracellular transport and signaling. Their elementary building blocks (globular proteins) self-assemble into cell-spanning filaments, and are strongly affected by temperature, ionic concentrations, pH, interacting proteins and other co-factors. The generated body of experimental data on the effects of environmental conditions (temperature, pH, ionic concentrations, electric and EM fields) on MTs and AFs, is available only to a very limited and fragmentary extent. Additionally, the dynamics of tubulin dimers and their assembly into MT structures (and similarly actin and AFs) is critical for many cellular processes, hence the modeling of the effects of EM interactions with the inclusion of cytoskeleton's dynamics is of great importance to understand how living cells work. This aspect involves details of these proteins' cyclical polymerization/depolymerization processes, which are not well understood as functions of temperature, ionic concentration and pH with simultaneous exposure to electric and EM fields. Interestingly, photobiomodulational effects have been reasonably well documented at tissue, organ and organism levels while mechanistic understanding of these effects at the cell level or below is almost completely absent. In particular, photobiological factors involve EM fields from the lower part of the ultraviolet (UV; approx.10 nm wavelength) to the upper part of the infrared (IR; approx.1 mm wavelength) of the EM spectrum. The interconnections between the subcellular elements (MTs, AFs and MAPs) can be controllably used as units of complex electrical circuits. A relevant manifestation of cell metabolism is the production of electron excitations in biomolecules leading to ultra-weak photon emission [29, 30], which is detectable in all metabolically active biological species. The EM activity of living cells involves a broad range of frequencies, from kHz to THz to IR, visible and even UV ranges. These frequency ranges (see Fig. 2) are related to electromechanical vibrations of subcellular structures. Our hypothesis is that EM fields significantly contribute to biological responses and hence to both intra- and inter-cellular communication.

Recently, their electrical conduction properties (see below) have implicated them as potential EM energy transducers. Being able to reverse engineer the complexity of intra-cellular interactions with electric and EM fields will help us better understand the basic principles of cell biology from the biophysics and bioengineering perspective and enable precise control of such

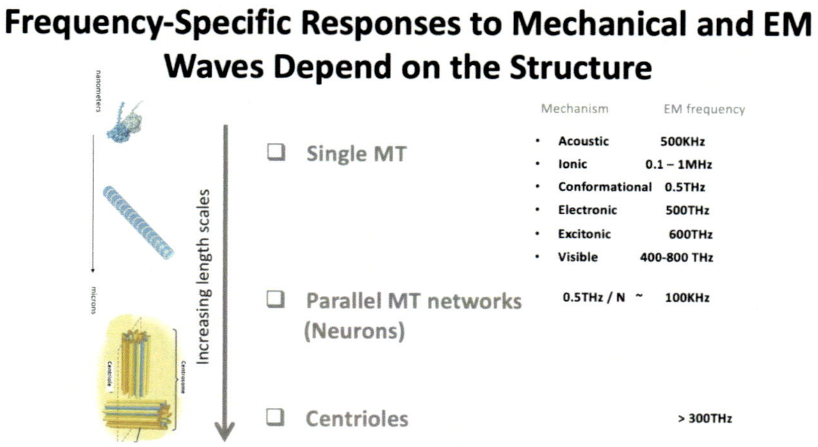

FIGURE 2
A summary of known and predicted characteristic resonant frequency ranges for the various tubulin-based structures

processes as cell differentiation and cell division by physical means, possibly in combination with the better-understood pharmacological agents.

DNA is a highly charged polymer whose folding depends on the electrostatic forces modulated by ions and pH [16]. AFs have been shown to send electrical signals losslessly over intra-cellular distances [17]. MTs exhibit a plethora of complicated electrical properties from capacitive [18], to highly conductive [11], to memristive [19] to transistor-like [38] signatures, depending on the ambient conditions. In Fig. 3 we show an experimentally determined memristive behavior of ionic conduction of MTs [19].

MTs polymerize at different rates depending on the presence/absence of EM fields acting on them. Membranes are well-known as leaky capacitors and hence are represented as RC electrical elements. However, there has been no comprehensive effort to date to construct an integrated circuit model for the entire cell similar to the mechanical tensegrity model of the cell. This is partly due to a lack of comprehensive understanding of the electrical properties of these key cellular structures. Recently, based on the earlier theoretical and computational models of MTs and AFs [8, 9, 10, 22], measurements [11, 18, 20, 21], and modelling, the following properties of tubulin and MTs were made: (a) hydration shell around tubulin that ranges from approx. 1 nm to several nm under reduced ionic concentrations, reduced pH or increased solvent concentration. This indicates that the environment controllably tunes up electrical and EM properties of tubulin and MTs. (b) The net electric charge of tubulin depends on the environment such that tubulin, which is normally highly negative, becomes positive when pH is below 5. Importantly, this positively-charged tubulin can still

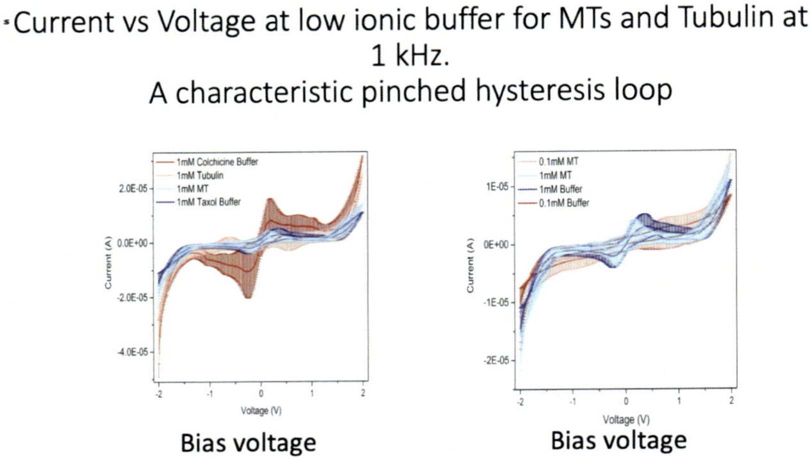

FIGURE 3
An illustration of memristive properties of microtubules [19].

polymerize into MTs. This was corroborated by electrophoretic mobility and zeta potential measurements. (c) Polarizability and the index of refraction of tubulin have been precisely measured and computed as a function of pH such that a varies in the range (4 to 9) x 10^{-34} Cm^2/V, for pH between 6.6 and 7.4 [39]. (d) The dielectric constant in the same range of pH varies between 2 and 4, which is much lower than previously reported in the literature [40]. (e) The real and imaginary parts of impedance of MTs and their ensembles were measured for three concentrations of tubulin between 0.222 μM and 22.2 μM and two ionic concentrations between 4 and 120 nM. At low ionic concentrations, the Debye length for MTs reaches tens of nm with MTs behaving as conductors with a conductivity 2-3 orders of magnitude greater than that of the surrounding solution. Conversely, at high ionic concentrations, MT's Debye lengths are on the order of 1 nm and MTs behave as capacitors (charge storage devices for counter ions), exhbiting conductive properties comparable to the ionic solution. The crossover point for the conductive-to-capacitive property of MTs in comparison to the solution is close to 100 mM. This indicates that in cellular environments, the roles of MTs can drastically change depending on the ionic concentrations of the cell (see Fig. 4). Hence, as electrical circuit elements, MTs can be represented as an R-C and sometimes L (see Fig. 5) combinations with frequency- and ionic concentration-dependent nonlinear electrical characteristics. This dependence on frequency and ionic composition is very important in cellular situations where conditions change. MTs engage in diverse interactions with ions and ionic currents, i.e.: (a) form a physical barrier to ionic conduction; (b) attract and accelerate ions along their axis

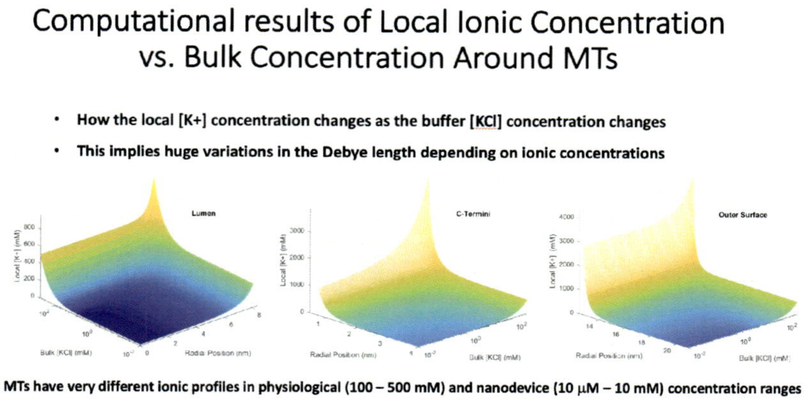

FIGURE 4
Computational predictions of the ionic charge distribution around microtubules as a function of distance and bulk concentration [22]

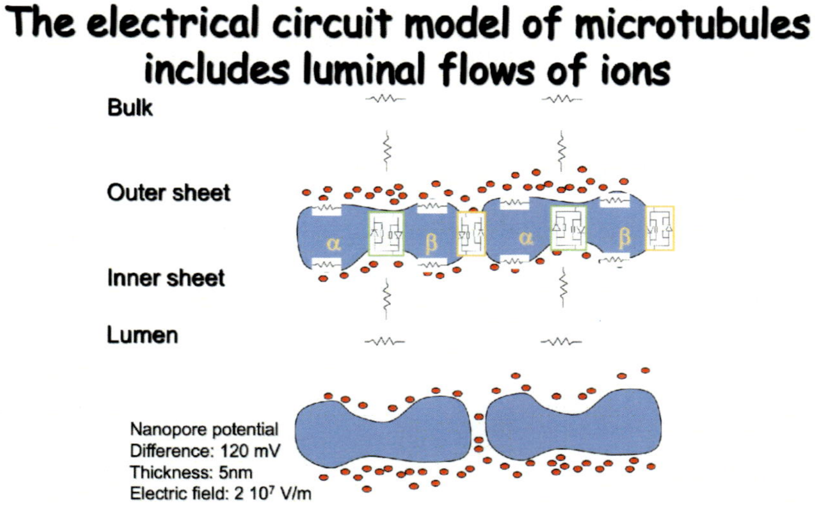

FIGURE 5
A graphical representation of ionic conduction along microtubule surfaces

leading to increased conductance; (c) condense ions on their surface decreasing mobile charge carriers concentration in the medium; (d) their C-termini create ion fluxes in and out of the lumen providing helical conduction pathways for solenoid-like behavior of MTs and contributing to MT's memristive properties under special conditions.

A specially designed scattering technique allowed to continuously examine the effects of temperature cycling on MT assembly/disassembly processes. Low-coherence dynamic light scattering measurements in a closed, *in vitro* tubulin-MT system showed a thermal hysteresis loop [41], consistently with the coexistence of tubulin and MTs in a phase transition [23]. Further, assembly/disassembly cycle ceased after several successive runs, indicating protein aging. Potential implications on human physiology, including the origin of diseases of aging, such as Alzheimer's and cancer, are enormous but require detailed studies (e.g. via Raman scattering) to determine the molecular mechanism involved. Especially in the case of Alzheimer's the evidence of neuronal MT damage is clear, although primarily due to hyperphosphorylation of MAP tau. A contributing factor could be the decreased ability of old tubulin to form MTs. Moreover, the process of MT assembly was seen to be strongly dependent on the exposure to EM fields in the visible range indicating the existence of interactions between charged polymers such as MTs or AFs and EM fields [24]. The latter can hence control the stability and assembly rates of the former.

Much is known about electrical property changes in cancer cells compared to their normal counterparts (see Fig. 6).

At the next level of biological organization is the response of entire cells to EM fields. While much is known about electrical property changes in cancer cells compared to their normal counterparts [25], virtually none of this knowledge has been translated into therapeutically-beneficial interventions. Briefly, the transmembrane potential is significantly reduced, intracellular pH is increased while extracellular pH decreased, ion channels operate at different rates and ionic concentrations are shifted, e.g. cellular potassium is increased. Also, MT dynamics is changed and mitotic spindle morphology

Electrophysiological hallmarks of cancer cells:

	Normal cells	Cancer cells
pH_i	6,99 – 7,05	7,2 – 7,8
pH_e	7,35 – 7,45	6,2 – 6,8
	$pH_i < pH_e$	$pH_i > pH_e$

	Normal cells	Cancer cells
pH_m	≈ 7,8	> 8,3
MV_m [mV]	≈ -138	≈ -210**

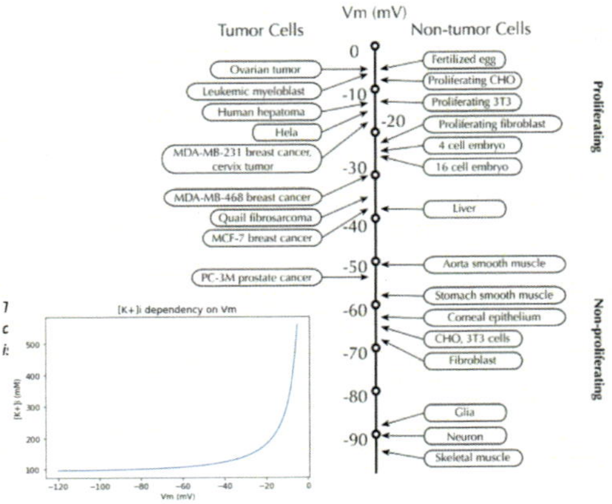

FIGURE 6
A graphical summary of the differences between cancer and normal cells in terms of the electrical properties [25].

altered. All of this should lead to changes in cellular responses to electrical fields. Yet, there is no integrated model that represents a quantitative response of a cancer cell to EM fields. Nonetheless, a therapeutic modality is under development that uses electric fields to target cancer cells, although its mode of action is still unclear. Specifically, Tumor Treating AC electric fields (TTFields) constitute a new FDA-approved modality to treat glioblastoma and mesothelioma. It is not known why specific frequencies (100-300 kHz) affect the proliferation of cancer cells. The effects of TTFields on MTs are of great interest due to their role as mitotic spindles in cell division. Thus, the improved electrical characterization of MTs is of major relevance to novel cancer therapeutics. I already initiated pilot studies of TTFields and EM effects on cancer cells where GFP-labeled tubulin expressed in HeLa cells was used to visualize the effects of TTFields on MTs in cancer cells showing that TTFields affect mitotic spindle morphology but a careful systematic study on other cell lines and under different conditions is much needed.

Somewhat similar to TTFields, photobiomodulation (PBM) therapy [26-28] is still lacking a mode of action and parameters selection for its clinical applications. PBM is based on the use of red or near-infrared (NIR) light to heal, restore, and stimulate physiological processes and to repair damage caused by injury or disease. However, under some conditions, it may promote tumour cell proliferation. It may be used in conjunction with other cancer therapies (e.g. radiation) to mitigate their side effects. PBM follows the "biphasic dose-response" (Arndt–Schulz) curve implicating the existence of ideal parameters for irradiated tissues, but exceeding these parameter values might cause harm. The wavelengths in the range 600-1000 nm show strong effects on cell proliferation. Shorter wavelengths are absorbed by haemoglobin or melanin, while longer ones by water, allowing light in this "optical window" to be absorbed by the cells. The energy of visible red-light photons is sufficient to increase reactive oxygen species (ROS) production which promotes growth factor synthesis, cell motility and viability. Greater energy density results in the inhibition of cell proliferation, migration, viability, and ATP activity [29, 30].

3 GAPS IN OUR KNOWLEDGE BASE

A recent review paper [10] outlined several key challenges in the field of bioelectricity. These challenges (see Table 1) address major gaps in our knowledge.

MT are distinct in their dynamic instability which includes both linear polymerization and stochastically distributed catastrophes. MTs are in fact hollow cylinders with $1nm$ pores on their surface allowing for ionic currents to flow in and out of the lumen. AF form branched networks in the presence of other proteins while MTs form regular lattices with interconnections pro-

Field of Impact	Scientific Question to be Answered
Intracellular biophysics	1) How do electric fields affect MT dynamic instability: MT nucleation, elongation, shortening rates, catastrophe frequency and rescue frequency?
Intracellular biophysics	2) Under which conditions MTs (and AFs) preferentially transport ions compared to the cytoplasm? Which ions are transported preferentially?
Intracellular biophysics	3) How do MAPs and drug binding, and the presence of AFs, alter MT electrical conductivity?
Intracellular biophysics	4) Does MT/tubulin polymerization (and MT orientation) alter the EM properties of the cell? Can they trigger intracellular events?
Intracellular biophysics	5) What are the ratios of ionic transport along MTs via: (a) lumen transport, (b) outer surface transport, (c) nanopore transport?
Intracellular biophysics +	6) How do MT-MAP interactions change under exposure to electric or EM fields? Do these fields cause biochemical changes in the cytoskeleton?
Nanodevice applications	7) How do MTs interact with acoustic waves? Can MT electrical properties be used for electro-mechanical coupling?
Nanodevice applications	8) How does ionic conductance depend on MT length and diameter? This is unclear due to the presence of nanopores and ionic 'leakage'
Intracellular biophysics + medical applications	9) What is the effect of cellular membranes on MT coupling to AC electric fields? How does the activity of ion channels influence MT coupling?

TABLE 1
A list of key challenges for MT bioelectricity.

vided by MAPs. MTs have a much higher electrostatic charge per length than AFs and their C-termini carry 40% of the charge making them antenna like objects.

MT assembly can be controlled by experimental conditions of temperature, ionic concentrations and pH. MTs can be stabilized/destabilized by pharmacological agents, e.g., taxane, colchicine, vinca alkaloids. By skilfully timing and dosing the administration of these compounds to the solution of tubulin in an appropriate buffer, assemblies of MTs can be generated with desirable length distributions. Using high concentrations of zinc added to these solutions, various structures can be generated in a controllable manner, e.g., 2D sheets of tubulin with antiparallel protofilament orientations and macrotubes which are cylinders made up of tubulin whose diameters are approximately 10 times greater than those of MTs. Such structures can be useful in characterizing capacitive, conductive and inductive properties of tubulin-based ionic conduction systems. Interconnections between individual

MTs can be easily created by mimicking natural solutions found in neurons, namely by adding microtubule-associated proteins (MAPs) to the MT-containing dishes. MAPs added there will form networks whose architecture can be determined by confocal microscopy and TEM imaging experiments. There is an almost inexhaustible range of possible architectures that can be built on the basis of MT and MAP assemblies. Their conductive properties are at the moment a completely unexplored area of research, which, based on what we know about MT conductive properties, is potentially a treasure trove of ionic conduction circuitry. Combining these circuits with actin-based circuits leads to a combinatorial explosion of possibilities that can only be described as a revolutionary transformation in the field of bioelectronics

4 METHODOLOGY

MT assembly can be controlled by experimental conditions of temperature, ionic concentrations and pH. MTs can be stabilized/destabilized by pharmacological agents, e.g. taxane, colchicine, vinca alkaloids. By skillfully timing and dosing the administration of these compounds to the solution of tubulin in an appropriate buffer, assemblies of MTs can be generated with desirable length distributions. Interconnections between MTs can be created by adding MAPs (e.g. MAP2 or tau) to the MT-containing dishes. MAPs added there form networks whose architecture can be determined by confocal microscopy. There is an almost inexhaustible range of possible architectures that can be built on the basis of MT and MAP assemblies. These supramolecular structures can be modeled at both the atomistic level (using MD simulations) and at a coarse-grained level (using COMSOL simulations), providing engineering blueprints for both cellular organization of the cytoskeleton and artificially created nano-bio-devices. Combining these circuits with actin-based circuits will lead to a combinatorial explosion of possibilities and lead to a revolutionary transformation in the field of bioelectronics. TTFields are most effective at 200 kHz while the electromechanical resonance of MTs due to the vibration of C-termini has been predicted to occur at 100 Hz. Beyond 1 MHz, ions are not responsive to alternating electric fields. However, as always, there exist certain technical challenges that need to be overcome in order to achieve tangible results.

The polymer chips can be made of biological polymers with ionic waves conductivity - actin filaments and tubulin microtubules, and synthetic polymers with solitonic and polaronic conductors: The polymer networks can have optical and electrical I/O and process information via interactions of travelling ionic condensations, dynamical changes of conductivity and electro-mechanical couplings. This can be implemented in a highly efficient manner as a high-performance massively-parallel computer. A range of computing primitives implemented by the polymer chip is determined and programmed by a geometry and graph-theoretic characteristics of the polymer network. Architectures of polymer computing chips can be grown via chemically controllable polymerisation reactions. The polymerisation is controlled and programmed by additional proteins to produce a stable structure, by electrical stimulation and by geometrical constraining of the physical space using patterned solid surfaces.

5 CONCLUSIONS

As argued above, a completely new and unexplored paradigm is emerging, namely that bioelectric properties of the cell and its intra- and extra-cellular environment can be controlled by external means in terms of electric fields, electromagnetic fields, temperature, pH and chemical agents. We need to open the curtain on the bio-electric engineering principles of the cell and its major components in order to reverse engineer the underlying design and exploit the knowledge of this design for controlling the state of the cell to our advantage. This could lead to the development of protein-based computational devices, nano-sensors or even entire (possibly evolvable) computers. We should do so by providing not only the experimentally-determined input data needed for the creation of an electrical circuit model of the cell but also to integrate these elements in a state-dependent, i.e. dynamical manner, so that cells in interphase should represent a different circuit compared to mitotic cells, an aspect addressed by the tensegrity theory for cell mechanics but not the cell electrodynamics. This is a complex and ambitious undertaking in view of the complexity of the cellular architecture, its state-dependent dynamics and the functional dependence of the electrical and EM properties of the cell and its components on numerous parameters, such as the frequency and intensity of the field, ionic concentration of the cytoplasm, pH and temperature. Through an iterative process of experimentation informing computational modeling, followed by modeling predictions being subjected to experimental validation and subsequent model refinement we can accomplish the overarching objective of providing a validated and robust model for entire cells in a state-dependent manner and separate models for normal and cancer cells with their characteristic values for the input parameters. A general schematic of such a model is shown in Fig. 7.

Electrical properties of AF and MT can be explored in terms of resistance, capacitance, transistor-like functionality, operational amplification, multiplication, fixed-function generation. Scoping devices should allow the execution of core mathematical operations such as: summation, integration with respect to time, inversion, multiplication, exponentiation, logarithm, and division. The mathematical problems are represented in polymer networks of resistive elements or reaction elements, with capacitance and inductance, spatial distribution of voltage, current, electrical potential in space, tempera-

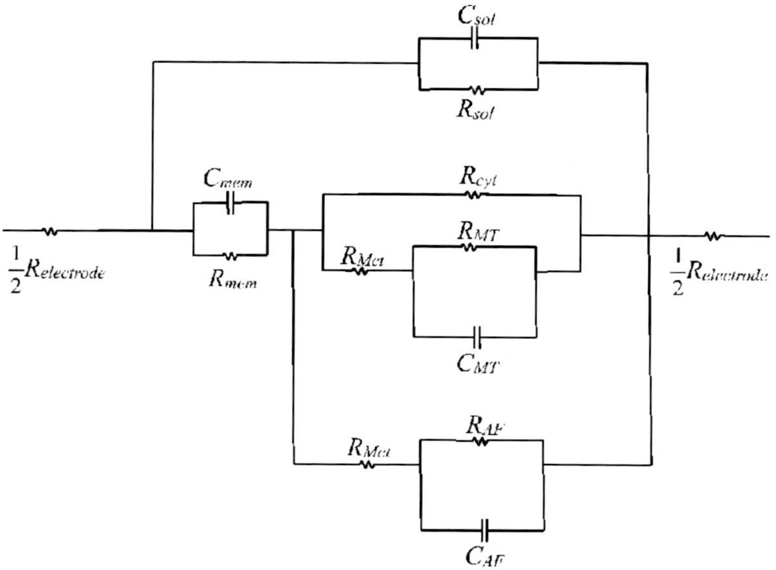

FIGURE 7
Schematic representation of the model representation of a living cell as a bioelectric device.

ture, pressure. Implementations could include active elements for the construction of analog computers, without amplification and passive computers, where amplification of signal is necessary.

Electrical I/O can be implemented via multi-electrode array technology [31] which gives us time resolution on the order of msec but offers the possibility to go towards device/chip configuration on top of which we can grow/deposit AT/MT bundles. These approaches might be augmented with nanofiber-light addressable potentiometric sensor [32] and stimulated-emission depletion microscopy (STED) combined with fluorescence correlation spectroscopy (FCS) with high spatial resolution (50-100 *nm*) and time resolution from tens of *nsec* to *msec* [33].

Developing Boolean circuits can also be implemented in polymer networks. The electrical properties of polymer networks can be used to implement logic gates using a flexible hardware platform designed for investigating the electrical properties of a substrate. One can apply arbitrary AC/DC signals to "configure" the network, i.e. to change its geometry, number of nodes and links, and connectivity, and then measure the network's electrical response. Complex sequences of digital signals and analogue waveforms can be generated and mapped arbitrarily to electrodes connected to the polymer network.

Biochemical properties such as tubulin's ligand binding sites and MT polymerization dynamics, are well researched but optical properties of tubulin and MTs have only recently gained interest due to the reported large dipole moment ≈ 2000 D and a charge of -50e of the tubulin heterodimer [18, 20 A40-42]. Extrapolating it to the entire cytoskeleton, composed of many thousands of such basic elements able to re-organize during the cell cycle on demand, the cell structure is expected to become a complex dynamical signal transduction device, which reconfigures itself depending on ambient conditions that can be controlled. Augmenting this with the existing information on the electrical properties of cell membranes and ion channels [34, 35, 36] will allow us to construct the first integrated computational representation of the cell as an evolvable electroconductive circuit device which also acts as an EM signal transducer propagating signals in a yet-to-be-revealed molecular mode of action.

The research area discussed here is expected to open a new frontier area in cell biophysics and nano-bio-technology where electric and EM effects are a natural part of the possible repertoire of responses, eventually combining it with pharmacology to add a new dimension to the range of modalities and capabilities. MTs, their complexes and AFs [36] can: (a) act as conduits for ionic signal propagation contributing to ligand-based signaling, (b) act as charge storage devices (equivalent to electrical batteries), (c) be used as memory storage devices (memristors) [19], (d) be used as EM signal transducers between the exterior environment and the cell, (e) send EM signals when mechanically excited since their large dipolar structures can oscillate and hence emit EM waves, (f) depolymerize when exposed to laser light in the visible range, (g) produce delayed luminescence effects in the visible range with half lives of the re-emitted radiation reaching second to minutes. Moreover, as R-L-C electrical circuits, they can be engaged in resonant interactions with a tunable range of frequencies that depend on: (a) pH and ionic composition of the cytoplasm or solution, (b) organization of MTs into supramolecular structures, (c) connections with other biomolecules such as MAPs. Systematically quantifying the above effects would represent a breakthrough in our understanding of a living cell, providing a design blueprint from a completely new EM perspective. This will lead to deep insights into how the basic elements of living systems work and how they can be controlled to our advantage in hybrid or redesigned configuratiuons. Transformational impact on science, technology and society is expected as future bioelectronic designs will be integrated at the nanoscale, where nano-templated growth of cytoskeleton polymer networks could form an architecture of conductive and information processing elements of intracellular circuits or as augmented hybrid devices. The impact of the research will also involve the ability to controllably design and utilize the building blocks for future nano-bio-sensors and electronic devices, which could be evolutionary, scaled-down versions of today's smart devices but at a nanoscale using cellular components. Research on nano-devices is aimed at learning the physical properties of very small structures and then determining how these can be used to perform computing, storage or reporting functions. The polymer chips could be made of biological polymers with ionic storage and conductivity properties, mainly MTs and MAPs but also possibly AFs These polymer networks could have EM (optical) and electrical I/O and process information *via* interactions involving ionic voltage signals, and electro-mechanical couplings in a highly efficient manner as massively-parallel signal transduction units. The development of polymer networks as nano-bio-devices could represent a major paradigm shift for future technology opening new avenues for medical implants, nanobiology, nano-neuroscience and nano-computing.

REFERENCES

[1] Maccioni, R.B. and Cambiazo, V., 1995. Role of microtubule-associated proteins in the control of microtubule assembly. *Physiological reviews*, 75(4), pp.835-864.

[2] Havelka, D., Cifra, M. and Kučera, O., 2014. Multi-mode electro-mechanical vibrations of a microtubule: in silico demonstration of electric pulse moving along a microtubule. *Applied Physics Letters*, 104(24).

[3] Havelka, D., Kučera, O., Deriu, M.A. and Cifra, M., 2014. Electro-acoustic behavior of the mitotic spindle: a semi-classical coarse-grained model. *PLoS One*, 9(1), p.e86501.

[4] Wenger, C., Salvador, R., Basser, P.J. and Miranda, P.C., 2015. The electric field distribution in the brain during TTFields therapy and its dependence on tissue dielectric properties and anatomy: a computational study. *Physics in Medicine & Biology*, 60(18), p.7339.

[5] Kirson, E.D., Giladi, M., Gurvich, Z., Itzhaki, A., Mordechovich, D., Schneiderman, R.S., Wasserman, Y., Ryffel, B., Goldsher, D. and Palti, Y., 2009. Alternating electric fields (TTFields) inhibit metastatic spread of solid tumors to the lungs. *Clinical & experimental metastasis*, 26, pp.633-640.

[6] Tuszynski, J.A., Wenger, C., Friesen, D.E. and Preto, J., 2016. An overview of sub-cellular mechanisms involved in the action of TTFields. *International journal of environmental research and public health*, 13(11), p.1128.

[7] Shrode, L.D., Tapper, H. and Grinstein, S., 1997. Role of intracellular pH in proliferation, transformation, and apoptosis. *Journal of bioenergetics and biomembranes*, 29, pp.393-399.

[8] Tuszyński, J.A., Brown, J.A., Crawford, E., Carpenter, E.J., Nip, M.L.A., Dixon, J.M. and Satarić, M.V., 2005. Molecular dynamics simulations of tubulin structure and calculations of electrostatic properties of microtubules. *Mathematical and computer modelling*, 41(10), pp.1055-1070.

[9] Tuszynski, J.A., Luchko, T., Carpenter, E.J. and Crawford, E., 2004. Results of molecular dynamics computations of the structural and electrostatic properties of tubulin and their consequences for microtubules. *Journal of Computational and Theoretical Nanoscience*, 1(4), pp.392-397.

[10] Kalra, A.P., Eakins, B.B., Patel, S.D., Ciniero, G., Rezania, V., Shankar, K. and Tuszynski, J.A., 2020. All wired up: an exploration of the electrical properties of microtubules and tubulin. *ACS nano*, 14(12), pp.16301-16320.

[11] Santelices, I.B., Friesen, D.E., Bell, C., Hough, C.M., Xiao, J., Kalra, A., Kar, P., Freedman, H., Rezania, V., Lewis, J.D. and Shankar, K., 2017. Response to alternating electric fields of tubulin dimers and microtubule ensembles in electrolytic solutions. *Scientific reports*, 7(1), p.9594.

[12] Albrecht-Buehler, G., 2005. A long-range attraction between aggregating 3T3 cells mediated by near-infrared light scattering. *Proceedings of the National Academy of Sciences*, 102(14), pp.5050-5055.

[13] Fels, D., 2009. Cellular communication through light. *PLoS one*, 4(4), p.e5086.

[14] Alberts, B., Bray, D., Hopkin, K., Johnson, A.D., Lewis, J., Raff, M., Roberts, K. and Walter, P., 2015. *Essential cell biology*. Garland Science.

[15] Preto, J., Pettini, M. and Tuszynski, J.A., 2015. Possible role of electrodynamic interactions in long-distance biomolecular recognition. *Physical Review E*, 91(5), p.052710.

[16] Gelbart, W.M., Bruinsma, R.F., Pincus, P.A. and Parsegian, V.A., 2000. DNA-inspired electrostatics. *Physics today*, 53(9), pp.38-45.

[17] Tuszyński, J.A., Portet, S., Dixon, J.M., Luxford, C. and Cantiello, H.F., 2004. Ionic wave propagation along actin filaments. *Biophysical journal*, 86(4), pp.1890-1903.

[18] Kalra, A.P., Patel, S.D., Bhuiyan, A.F., Preto, J., Scheuer, K.G., Mohammed, U., Lewis, J.D., Rezania, V., Shankar, K. and Tuszynski, J.A., 2020. Investigation of the electrical properties of microtubule ensembles under cell-like conditions. *Nanomaterials*, 10(2), p.265.

[19] Tuszynski, J.A., Friesen, D., Freedman, H., Sbitnev, V.I., Kim, H., Santelices, I., Kalra, A.P., Patel, S.D., Shankar, K. and Chua, L.O., 2020. Microtubules as sub-cellular memristors. *Scientific reports*, 10(1), p.2108.

[20] Kalra, A.P., Patel, S.D., Eakins, B.B., Riddell, S., Kumar, P., Winter, P., Preto, J., Carlson, K.W., Lewis, J.D., Rezania, V. and Tuszyński, J.A., 2021. Revealing and attenuating the electrostatic properties of tubulin and its polymers. *Small*, 17(1), p.2003560.

[21] Kalra, A.P., Kar, P., Preto, J., Rezania, V., Dogariu, A., Lewis, J.D., Tuszynski, J.A. and Shankar, K., 2019. Behavior of α, β tubulin in DMSO-containing electrolytes. *Nanoscale Advances*, 1(9), pp.3364-3371.

[22] Eakins, B.B., Patel, S.D., Kalra, A.P., Rezania, V., Shankar, K. and Tuszynski, J.A., 2021. Modeling microtubule counterion distributions and conductivity using the Poisson-Boltzmann equation. *Frontiers in Molecular Biosciences*, 8, p.650757.

[23] Sept, D. and Tuszyńskit, J.A., 2000. A Landau-Ginzburg model of the Co-existence of free tubulin and assembled microtubules in nucleation and oscillations phenomena. *Journal of Biological Physics*, 26, pp.5-15.

[24] Farhi, A. and Dogariu, A., 2021. Coupling of electrodynamic fields to vibrational modes in helical structures. *Physical Review A*, 103(2), p.023523.

[25] Di Gregorio, E., Israel, S., Staelens, M., Tankel, G., Shankar, K. and Tuszyński, J.A., 2022. The distinguishing electrical properties of cancer cells. *Physics of life reviews*, 43, pp.139-188.

[26] Zein, R., Selting, W. and Hamblin, M.R., 2018. Review of light parameters and photobiomodulation efficacy: dive into complexity. *Journal of biomedical optics*, 23(12), pp.120901-120901.

[27] Del Vecchio, A., Tenore, G., Luzi, M.C., Palaia, G., Mohsen, A., Pergolini, D. and Romeo, U., 2021, January. Laser photobiomodulation (Pbm)—a possible new frontier for the treatment of oral cancer: A review of in vitro and in vivo studies. In *Healthcare* (Vol. 9, No. 2, p. 134). MDPI.

[28] Tam, S.Y., Tam, V.C., Ramkumar, S., Khaw, M.L., Law, H.K. and Lee, S.W., 2020. Review on the cellular mechanisms of low-level laser therapy use in oncology. *Frontiers in Oncology*, 10, p.1255.

[29] Devaraj, B., Usa, M. and Inaba, H., 1997. Biophotons: ultraweak light emission from living systems. *Current Opinion in Solid State and Materials Science*, 2(2), pp.188-193.

[30] Cifra, M., Brouder, C., Nerudová, M. and Ku era, O., 2015. Biophotons, coherence and photocount statistics: A critical review. *Journal of Luminescence*, 164, pp.38-51.

[31] Spira, M.E. and Hai, A., 2013. Multi-electrode array technologies for neuroscience and cardiology. *Nature nanotechnology*, 8(2), pp.83-94.

[32] Shaibani, P.M., Jiang, K., Haghighat, G., Hassanpourfard, M., Etayash, H., Naicker, S. and Thundat, T., 2016. The detection of Escherichia coli (E. coli) with the pH sensitive hydrogel nanofiber-light addressable potentiometric sensor (NF-LAPS). *Sensors and Actuators B: Chemical*, 226, pp.176-183.

[33] Lanzanò, L., Scipioni, L., Di Bona, M., Bianchini, P., Bizzarri, R., Cardarelli, F., Diaspro, A. and Vicidomini, G., 2017. Measurement of nanoscale three-dimensional diffusion in the interior of living cells by STED-FCS. *Nature communications*, 8(1), p.65.

[34] Pietak, A. and Levin, M., 2016. Exploring instructive physiological signaling with the bioelectric tissue simulation engine. *Frontiers in bioengineering and biotechnology*, 4, p.55.

[35] Churchill, C.D., Winter, P., Tuszynski, J.A. and Levin, M., 2019. EDEn–electroceutical design environment: ion channel tissue expression database with small molecule modulators. *Iscience*, 11, pp.42-56.

[36] Siccardi, S., Adamatzky, A., Tuszyński, J., Huber, F. and Schnauß, J., 2020. Actin networks voltage circuits. *Physical Review E*, 101(5), p.052314.

[37] Blackiston, D.J., McLaughlin, K.A. and Levin, M., 2009. Bioelectric controls of cell proliferation: ion channels, membrane voltage and the cell cycle. *Cell cycle*, 8(21), pp.3527-3536.

[38] Priel, A., Ramos, A.J., Tuszynski, J.A. and Cantiello, H.F., 2006. A biopolymer transistor: electrical amplification by microtubules. *Biophysical journal*, 90(12), pp.4639-4643.

[39] Guzman-Sepulveda, J.R., Wu, R., Kalra, A.P., Aminpour, M., Tuszynski, J.A. and Dogariu, A., 2019. Tubulin polarizability in aqueous suspensions. *ACS omega*, 4(5), pp.9144-9149.

[40] Minoura, I. and Muto, E., 2006. Dielectric measurement of individual microtubules using the electroorientation method. *Biophysical journal*, 90(10), pp.3739-3748.

[41] Wu, R., Guzman-Sepulveda, J.R., Kalra, A.P., Tuszynski, J.A. and Dogariu, A., 2022. Thermal hysteresis in microtubule assembly/disassembly dynamics: The aging-induced degradation of tubulin dimers. *Biochemistry and Biophysics Reports*, 29, p.101199.

On the Positive Role of Noise and Cross-Talk of Elements in Neuromorphic Systems

VICTOR EROKHIN

Institute of Materials for Electronics and Magnetism, Italian National Research Council (IMEM-CNR), Parco Area delle Scienze, 37A, Parma, 43124, Italy

Received: September 1, 2024. Accepted: September 28, 2024.

Roles of three important properties of the neuromorphic systems are considered in this paper: effect of noise, cross-talk of elements and energy efficiency. The positive effect of noise, directly demonstrated theoretically and experimentally during recent years, allows the system to reach a global energy level, providing effective classification and forecasting tasks execution. Cross-talk of elements is important for the realization of bio-realistic learning algorithms and mimicking cognitive and creative processes. Energy efficiency is an obvious important parameter for all modern systems. In this paper we underline the necessity of applying new hardware approaches for the effective solving of this problem.

Keywords: Neuromorphic systems, memristive devices, noise, cross-talk of elements, energy efficiency

1 INTRODUCTION

The term "neuromorphic" began to be used during the end of 1980ies [1-3], beginning of 1990ies [4, 5], and first part of 2000ies [6]. The interest in such systems was significantly increased after the publication of 2008 [7], where it was claimed that a theoretically predicted memristor [8] was physically realized. The fact that this device was considered to have properties similar to biological synapses, the number of publications, dedicated to neuromorphic systems, was exponentially increased [9-18].

*Corresponding author email: victor.erokhin@imem.cnr.it

In this paper I will consider only hardware implementations of neuromorphic systems, because its software implementations are widely discussed in literature and they require a very high energy consumption (will be discussed later). Moreover, this paper will be mainly dedicated to hardware systems, realized using memristive devices (or resistive switching elements [19-22]), as they are considered now as most promising elements for the realization of electronic synapse analogs and essential components of artificial neurons.

It is obvious that neuromorphic systems are considered a step forward regarding the next generation of computers. However, different research groups imply different meanings to this term. In general, the realized system is considered as a "neuromorphic" one if it can mimic at least one essential property of the brain (nervous system). From my point of view, it is not completely correct: the system must have several essential properties, allowing us to consider it as a neuromorphic one.

In my previous works I have identified essential properties of neuromorphic systems [23, 24]. Some of them are absolutely supported by all researchers working in this area. For example, these systems must use the same elements for the storage and processing of the information, as it occurs in the brain. This branch is now accepted both in the software and hardware communities and is considered as a new paradigm, called "in memory computing" [25-27], that should accelerate speed of the computing (the exchange rate of the information between the memory and the processor is one of the slowest process at the software level) and significantly improve the energy efficiency (at the hardware level).

The other essential feature is the capability to mimic important features of the brain and nervous system, such as learning and adaptations. The necessity of having these properties for neuromorphic systems is also accepted by all research groups working in this area. In particular, bio-plausible STDP (Spike-Timing Dependent Plasticity) algorithm has been demonstrated in artificial systems, based on memristive devices [28-31]. I want to underline here that this algorithm is responsible for the establishing of automatic causal relations during unsupervised learning of all living beings. Moreover, it has been demonstrated the possibility of classic conditioning (Pavlov's dog learning) based on STDP algorithm in systems, based on organic memristive devices [32].

The other requirement, also accepted by the society, is the possibility to be integrated with elements of the nervous system [33].

The other item is related to the stochastic 3D organization of the network, allowing the formation of multiple signal pathways in it [34]. There are not so many publications regarding such kinds of organization [35-37]. However, this approach can establish a basis for the hardware realization of "reservoir computing" [38-40], that, at least basing on memristive devices, up to now was represented by simplified electric systems using short-term memory properties of memristive devices, organized in a deterministic circuits [41-46].

However, there are two items that were identified in my previous works [23, 24], namely, the positive role of the noise and necessity of having cross-talk of elements, turned out to be not so obvious to the researchers working in the field.

Thus, in this paper I want to clarify my opinion regarding these items. In addition, I will briefly mention another essential feature of the system, allowing us to consider it as a neuromorphic one – energy efficiency. I have not considered this item in my previous works, but it seems very important, especially taking into account forecasted difficulties, related to the use of traditional sources (multi-nuclei computer cards or internet Artificial Intelligence (AI) classification/identification/prediction).

2 POSITIVE ROLE OF NOISE

In traditional electronic devices and systems, the role of noise is practically always negative, as it is responsible for the decreased reliability of circuit performances. As a result, special shielding is usually used for the protection.

In the case of living beings, the situation is different: the noise can play a positive role [47, 48].

Obviously, as neuromorphic systems must mimic essential features of living beings, the noise must play a similar positive role also for them.

The schematic of the possible positive role of the noise is shown in Fig. 1.

Let us imagine a system having several energy minima. The system must classify events (randomly falling down spheres in the Fig. 1a). If there is no noise in the system, the spheres will be randomly distributed and some of them will even remain also outside the energy minima on the flat surfaces of the system (Fig. 1b). When the noise (vertical and horizontal vibrations of the system) of rather small amplitude will be applied to the system, the final distribution of spheres will be shifted to energy minima (no spheres will be found on flat surfaces), as it is shown in Fig. 1c. Further increase of the noise level amplitude will result in the localization of spheres only in global energy minima levels, as it is shown in Fig. 1d. Of course, if the noise level will be very high (the vibration amplitude will be higher than the depths of global energy minima levels) spheres again will be randomly distributed in the system. Thus, there is an optimal level of noise that can improve the performance of a particular system for solving a specific task.

Regarding neuromorphic systems based on memristive devices, the first theoretical mention of the possible positive effect of noise was done in 2012 [49]. The conclusions, done by authors, were independently experimentally confirmed by the other research groups [50, 51]. After it, several publications have also demonstrated the positive role of noise for the memristive-based neuromorphic systems [52, 53].

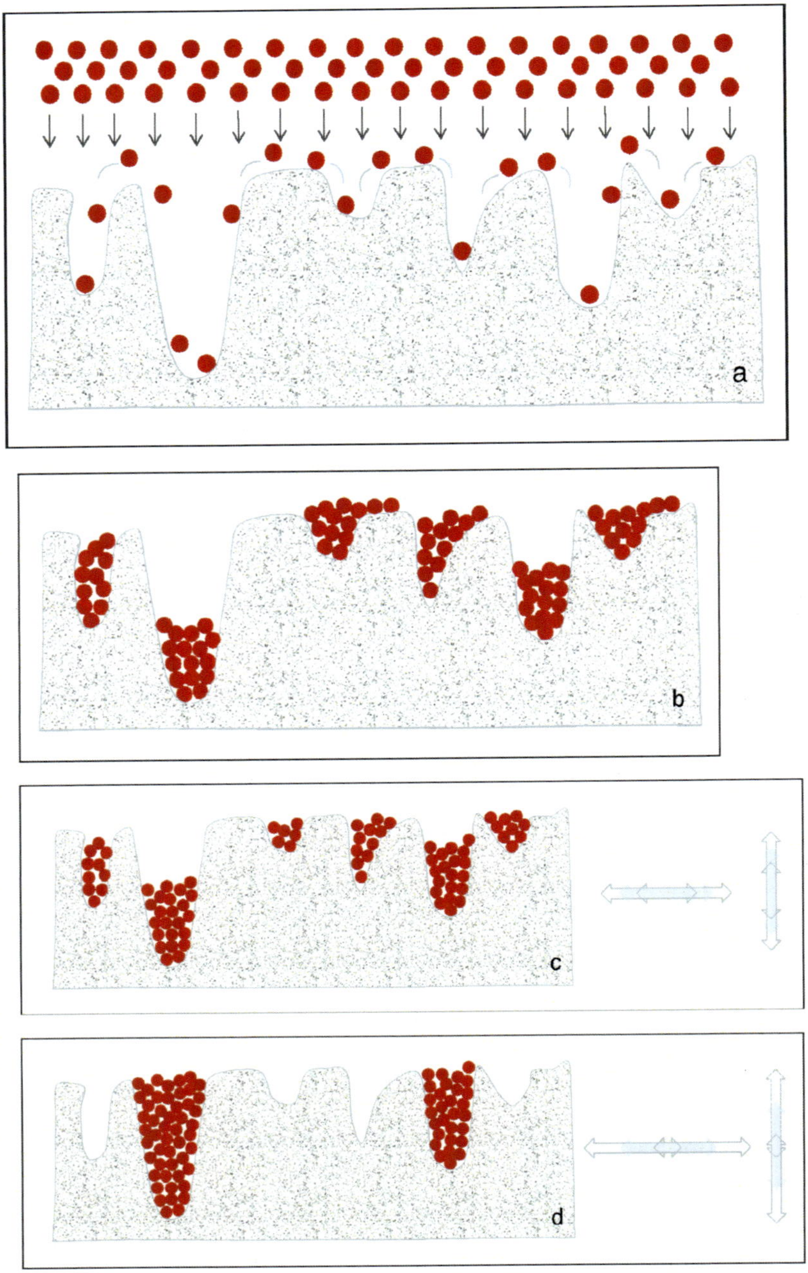

FIGURE 1
Schematic illustration of the positive role of noise in the system, having different energetic minima. Red spheres are events. a. Events are applied to the system with different energy minima; b. events are distributed in the system, when no noise is applied; c. low amplitude noise is applied to the system: events are located in energy minima (not on flat surfaces); d. high amplitude noise is applied to the system: events are located only in global energy minima.

The real breakthrough in the demonstration of the positive role of noise for neuromorphic information processing was done during the last five years [54-58]. Now the importance of the noise is obvious for all researchers, working in the field of hardware implementation of neuromorphic systems.

To conclude this section, I want to underline that the noise plays absolutely different roles in traditional electronic circuits and neuromorphic systems. In the first case it always performs a negative effect, while in the second case its role is important. Moreover, in my opinion, it must be seriously taken into account for mimicking cognitive and creative processes. In fact, the presence of noise can bring the neuromorphic system to a global energy minimum (as it is shown in Fig. 1), which can be considered as a creation of new knowledge, not possible in the noise-free conditions.

3 CROSS-TALK OF ELEMENTS

The other important item of neuromorphic systems is the role of cross-talk of their elements. Similarly to the case of the noise role, it is considered as an absolutely negative feature in the case of traditional electronic circuits. It is to note, that even considering quasi-neuromorphic systems, based on memristive devices, most researchers consider it as a negative one and are trying to delete or minimize this effect [59-65].

There are not so many papers dedicated to a positive role of the cross-talk between elements, even for systems, based on memristive devices [66-68].

In [66] it is shown that cross-talk of artificial neurons, linked through synaptic (memristive) connection can trigger multi-synchrony states of the system. Together with the presence of noise within some range it can result in the appearance of coherent neuronal patterns.

In [67, 68] it has been shown that cross-talk of memristive elements results in the formation of stable chains of the signal transfer, when several synapse-like elements are involved. It is possible to make a parallel to the brain function. The appearance of epilepsy, for example, is usually connected to the disbalances of connections between neurons. As it was shown in memristive-based artificial systems [67, 68], a stable chain of the information propagation can be reached only if resistances of all elements, establishing this chain (memristive devices), are balanced. In normally functioning system, the decrease of the conductivity of a certain device in the signal transferring chain (caused by fluctuation or external stimuli action) will be compensated, because the potential profile of the voltage, applied to the chain terminals, will be re-distributed in such a way, that the main part of the applied voltage will be located on this particular element, increasing its conductivity and decreasing slightly the conductivity of all other adjustment elements.

It is possible to claim that in natural neuron networks (nervous system) the cross-talk of elements is not only essential, but even a required function.

Moreover, it seems that the strength of the cross-talk (mutual influence of one to the other) must depend on the distance between "talking" elements.

For the supporting the previous statement, I want to refer to the paper [69], where it has been realized a calculated model of the system, composed of threshold nodes (neurons) distributed in 3D matrix through synapse-like (memristive) connections. In this model two types of threshold elements connections were considered: random distribution of the connections mode and distance dependent probability of connections. The results, obtained during this modelling, have been applied for the interpretation of the previously obtained learning capabilities, registered in 3D networks with stochastically distributed synaptic connections, based on organic memristive devices, obtained by self-organized approach [70]. It has been shown that experimentally observed dependences can be better explained by the model system, where the distant dependent connections were used. Moreover, at the level of the model, such a system has demonstrated the ability to unsupervised learning [69].

4 ENERGY EFFICIENCY

We are living in the epoch when the energy efficiency is considered as one of the most (or even the most) important parameter for practically devices, fabricated by humankind, that must be considered also (or even with more attention) for the realization of systems with "smart" properties, allowing classification, learning and forecasting, using computers which role is exponentially increasing during last years. Multi-nuclei cards, designed for the analysis of big data, require huge energy expenses. Detailed consideration of the energy consumption in different software and hardware-realized systems, with AI properties, are considered in [71].

Software realization of searching engines as well as AI classification and forecasting systems require enormous energy consumption. Thus, powerful cooling systems are usually associated with them, and these systems can consume even more energy than it is required for the AI task execution. It is expected that energy consumption only for the execution of AI-related problems will reach in 2027 the value within the range 85 – 134 TWh [72].

This number is very impressive. It is comparable to the total energy consumed by the whole humankind for carrying out mental processes. Thus, the application of traditional computers and solving of AI tasks at the software level can deepen the energetic crisis, widely discussed now.

According to me, the effective solution of this problem is related to the realization of new hardware neuromorphic architectures of unconventional computers, where each event will have the energy consumption, comparable to the energy, necessary to the energy of the synapse weight adjustment in nervous systems.

Currently, memristive devices are considered as promising key elements for such systems. The most attention is dedicated to their inorganic realization because it can be done using the existing technological tools and can be integrated with well-developed CMOS devices [73]. However, regarding the power consumption these devices have some drawbacks. Most such devices require the forming processes, where several volts must be applied to the system, resulting in several mA current. Thus, one event will require about 0.1 – 1.0 W per junction. Considering that the switching time is usually within the range of microseconds, we have 100 – 1000 nJ for the junction. Of course, if we consider a system comparable with the complexity of the brain (10^{15} synapses), the energy efficiency is very low.

The other situation can be observed in organic neuromorphic systems. Here I want to refer to only two examples. The first one is connected to the device I am working with: polyaniline based memristive device [24]. In this case, the min energy consumption for the practically implemented device was about 16 nJ per event [74]. It is to note that this value can be significantly improved if the size of the device will be reduced to submicron level.

The other example of an extremely energy efficient system is electrochemical neuromorphic organic device [75]. The energy consumption in this case was less than 10 pJ per one switching event, which is comparable with the energy consumption in the nervous system.

5 CONCLUSIONS

In this short paper I tried to underline important properties that neuromorphic systems must have. First, a noise must have a positive role for helping the system arrive at a more stable state, corresponding to the global energy minimum. Second, cross-talk of elements of the system must provide bio-realistic learning algorithms, such as STDP, and establish a basis of cognitive and creative processes. Finally, energy efficiency is also a key parameter regarding, in particular, AI and bio-mimicking applications. According to me, this item can be successfully addressed only at the hardware level, using unconventional computing approaches.

ACKNOWLEDGEMENT

I want to thank Prof. A Adamatzky for the organization of Gagarin's readings conference, where these results were presented and critically discussed with colleagues. I also want to thank Dr. Svetlana Erokhina for the help in the figure preparation and useful discussions.

REFERENCES

[1] A. Guez and J. Selinsky, "A trainable neuromorphic controller", J. Robotic Systems, 5, 363-388 (1988).

[2] N.H. Farhat and B.C. Bai, "Echo inversion and target shape estimation by neuromorphic processing", Neural Networks, 2, 117-125 (1989).

[3] E. Wacholder, J. Han, and R.C. Mann, "A neural Network algorithm for the multiple traveling Salesman problem", Biol. Cybernetics, 61, 11-19 (1989).

[4] C. Mead, "Neuromorphic electronic systems", Proc. IEEE, 78, 1629-1636 (1990).

[5] R. Douglas, M. Mahowald, and C. Mead, "Neuromorphic analog VLSI", Annual Rev. Neurosci., 18, 255-281 (1995).

[6] G. Indiveri, E. Chicca, and R. Douglas, "A VLSI array of low-power spiking neurons and bistable synapses with spike-timing dependent plasticity", IEEE Trans. Neuron Networks., 17, 211-221 (2006).

[7] D.B. Strukov, G.S. Snider, D.R. Stewart, and R.S. Williams, "The missing memristor found", Nature, 453, 80-83 (2008).

[8] L.O. Chua, "memristor – the missing circuit element", IEEE Trans. Circuit Theor., 18, 507-519 (1971).

[9] S.H. Jo, T. Chang, I. Ebong, B.B. Bhadvivya, P. Mazumder, and W. Lu, "Nanoscale memristor device as synapse in neuromorphic systems", Nano Lett., 10, 1297-1301 (2010).

[10] J.J. Yang, D.B. Strukov, and D.R. Stewart, "Memristive devices for computing", Nature Nanotechnol., 8, 13-24 (2013).

[11] M. Prezioso, F. Merrikh-Bayat, B.D. Hoskins, G.C. Adam, K.K. Likharev, and D.B. Strukov, "Training and operation of an integrated neuromorphic network based on metal-oxide memristors", Nature, 521, 61-64 (2015).

[12] Z.R. Wang, S. Joshi, S.E. Savel'ev, H. Jiang, R. Midya, P. Lin, M. Hu, N. Ge, J.P. Strachan, Z.Y. Li, Q. Wu, M. Barne, G.L. Li, H.L. Xin, R.S. Williams, and J.J. Yang, "Memristors with diffusive dynamics as synapse emulators for neuromorphic computing", Nature Materials, 16, 101-108 (2017).

[13] M. Davies, N. Srinivasa, T.H. Lin, G. Chinya, Y.Q. Cao, S.H. Choday, G. Dimou, P. Joshi, N. Imam, S. Jain, Y.Y. Liao, C.K. Lin, A. Lines, R.K. Paul, J. Tse, G. Venkataramanen, Y.H. Weng, A. Wild, Y. Yang, and H. Wang, "Loihi: A neuromorphic manycore processor with on-chip learning", IEEE Micro, 38, 82-99 (2018).

[14] P. Yao, H.Q. Wu, B. Gao, J.S. Tang, Q.T. Zhang, W.Q. Zhang, J.J. Yang, and H. Qian, "Fully hardware-implemented memristor convolutional neural network", Nature, 577, 641-646 (2020).

[15] Z.R. Wang, H.Q. Wu, G.W. Burr, C.S. Hwang, K.L. Wang, Q.F. Xia, and J.J. Yang, "Resistive switching materials for information processing", Nature Rev. Mater., 5, 173-195 (2020).

[16] J. Grollier, D. Querlioz, K.Y. Camsari, K. Everschor-Sitte, S. Fukami, and M.D. Stiles, "Neuromorphic spintronics", Nature Electronics, 3, 360-370 (2020).

[17] C.D. Schuman, S.R. Kulkarni, M. Parsa, J.P. Mitchell, P. Date, and B. Kay, "Opportunities for neuromorphic computing algorithms and applications", Nature Computational Sci., 2, 10-19 (2022).

[18] S. Kumar, X.X. Wang, J.P. Strachan, Y.C. Yang, and W.D. Lu, "Dynamical memristors for higher-complexity neuromorphic computing", Nature Rev. Mater., 7, 575-591 (2022).

[19] H. Pagnia and N. Sotnik, "Bistable switching in electroformed metal-insulator-metal devices", Phys. Status Solidi, 108, 11-65 (1988).

[20] A. Asamitsu, Y. Tomioka, H. Kuwahara, and Y. Tokura, "Current switching of resistive states in magnetoresistive manganites", Nature, 388, 50-52 (1997).

[21] R. Waser and M. Aono, "Nanobionics-based resistive switching", Nature Mater., 6, 833-840 (2007).

[22] M. Janousch, G.I., Meijer, U. Staub, B. Delley, S.F., Karg, and B.P. Andreasson, "Role of oxygen vacancies in Cr-doped ScTiO3 for resistance-change memory", Adv. Mater., 19, 2232-2235 (2007).

[23] V. Erokhin, "Memristive devices for neuromorphic applications: comparative analysis", BioNanoScience, 10, 834-847 (2020).

[24] V. Erokhin, "Fundamentals of organic neuromorphic systems", Springer Nature Switzerland AG, Switzerland (2022).

[25] M.A. Zidan, J.P. Strachan, and W.D. Lu, "The future of electronics based on memristive systems", Nature Electronics, 1, 22-29 (2018).

[26] D. Ielmini and H.S.P. Wong, "In-memory computing with resistive switching devices", Nature Electronics, 1, 333-343 (2018).

[27] Q.F. Xia and J.J. Yang, "Memristive crossbar arrays for brain-inspired computing", Nature Materials, 18, 309-323 (2019).

[28] C. Zamarreno-Ramos, L.A. Camunas-Mese, J.A. Perez-Carrasco, T. Masquelier, T. Serrano-Gotarredona, and B. Linares-Barranco, "On spike-timing-dependent-plasticity, memristive devices, and building a self-learning visual cortex", Frontiers Neurosci., 5, 26 (2011).

[29] S. Ambrosio, N. Ciocchini, M. Laudato, V. Milo, A. Pirovano, P. Fantini, and D. Ielmini, "Unsupervised learning by spike timing dependent plasticity in phase change memory (PCM) synapses", Frontiers Neurisci., 10, 56 (2016).

[30] N.V. Prudnikov, D.A. Lapkin, A.V. Emelyanov, A.A. Minnekhanov, Y.N. Malakhova, S.N. Chvalun, V.A. Demin, and V.V. Erokhin, "Associative STDP-like learning of neuromorphic circuits based on polyaniline microdevices", J. Phys. D. Appl. Phys., 53, 41401 (2020).

[31] V.A. Demin, D. Nekhaev, I.A. Surazhevsky, K.E. Nikiruy, A.V. Emelyanov, S.N. Nikolayev, V.V. Rylkov, and M. Kovalchuk, "Necessary conditions for STDP-based pattern recognition learning in a memristive spiking neural network", Neural Networks, 134, 64-75 (2021).

[32] A.A. Minnekhanov, A.V. Emelyanov, D.A. Lapkin, K.E. Nikiruy, B.S. Shvetsov, A.A. Nesmelov, V.V. Rylkov, V.A. Demin, and V.V. Erokhin, "Parylene based memristive devices with multilevel resistive switching for neuromorphic applications", Sci. Rep., 9, 10800 (2019).

[33] E. Jusekaeva, A. Nasreddinov, S. Battistoni, T. Berzina, S. Iannotta, R. Khazipov, V. Erokhin, and M. Mukhtarov, "Coupling cortical neurons through electronic memristive synapse", Adv. Mater. Technol., 4, 1800350 (2019).

[34] V. Erokhin, T. Berzina, K. Gorshkov, P. Camoreni, A. Pucci, L. Ricci, G. Ruggeri, R. Sigala, and A. Schuz, "Stochastic hybrid 3D matrix: learning and adaptation of electrical properties", J. Mater. Chem., 22, 22881-22887 (2012).

[35] V. Erokhin, T. Berzina, P. Camorani, and M.P. Fontana, "Conducting polymer-solid electrolyte fibrillar composite material for adaptive networks", Soft Matter, 2, 870-874 (2006).

[36] S. Erokhina, V. Sorokin, and V. Erokhin, "Skeleton-supported stochastic networks of organic memristive devices: adaptations and learning", AIP Adv., 5, 027129 (2015).

[37] Y.N. Malakhova, A.N. Korovin, D.A. Lapkin, S.N. Malakhov, V.S. Shcherban, E.B. Pichkur, S.N. Yakunin, V.A. Demin, S.N. Chvalun, and V. Erokhin, "Planar and 3D fibrous polyaniline-based materials for memristive elements", Soft Matter, 13, 7300-7306 (2017).

[38] D. Verstraeten, B. Schrauwen, M. D'Haene, and D. Stroobandt, "An experimental unification of reservoir computing methods", Neural Networks, 20, 391-403 (2007).

[39] M. Lukosevicius and H. Jaeger, "Reservoir computing approaches to recurrent neural network training", Computer Sci. Rev., 3, 127-149 (2009).

[40] G. Tanaka, T. Yamane, J.B. Heroux, R. Nakane, N. Kanazawa, S. Takeda, H. Numata, G. Nakano, and A. Hirose, "Recent advances in physical reservoir computing: a review", Neural Networks, 115, 100-123 (2019).

[41] C. Du, F.X. Cai, M.A. Zidan, W. Ma, S.H. Lee, and W.D. Lu, "Reservoir computing using dynamic memristors for temporal information processing", Nature Commun., 8, 2204 (2017).

[42] J. Moon, W. Ma, J.H. Shin, F.X. Cai, C. Du, S.H. Lee, and W.D. Lu, "Temporal data classification and forecasting using a memristor-based reservoir computing system", Nature Electronics, 2, 480-487 (2019).

[43] R. Midya, Z.R. Wang, S. Asapu, X.M. Zhang, M.Y. Rao, W.H. Song, Y. Zhuo, N. Upadhyay, Q.F. Xia, and J.J. Yang, "Reservoir computing using diffusive memristors", Adv. Intelligent Systems, 1, 1900084 (2019).

[44] Y.A. Zhong, J.S. Tang, X.Y. Li, B. Gao, H. Qian, and H.Q. Wu, "Dynamic memristor-based reservoir computing for high-efficiency temporal signal processing", Nature Commun., 12, 408 (2021).

[45] N.V. Prudnikov, V.A. Kulagin, S. Battistoni, V.A. Demin, V.V. Erokhin, and A.V. Emelyanov, "Polyaniline-based memristive devices as key elements of robust reservoir computing for image classification", Physica Status Solidi A, 2200700 (2023).

[46] A.N. Matsukatova, N.V. Prudnikov, V.A. Kulagin, S. Battistoni, A.A. Minnekhanov, A.D. Trofimov, A.A. Nesmelov, S.A. Zavyalov, Y.N. Malakhova, M. Parmiggiani, A. Balesio, S.L. Marasso, S.N. Chvalun, V.A. Demin, A.V. Emelyanov, and V. Erokhin, "Combination of organic-based reservoir computing and spiking neuromorphic systems for a robust and efficient pattern classification", Adv. Intelligent Systems, 5, 2200407 (2023).

[47] M.D. McDonnell and D. Abbott, "What is stochastic resonance? Definitions, misconceptions, debates, and its relevance to biology", PLOS Comput. Biol., 5, e1000348 (2009).

[48] F.Q. Wu, D.J. men, and X. Wang, "Quorum-sensing crosstalk-driven synthetic circuits: from unimodality to trimodality", Chem. Biol., 21, 1629-1638 (2014).

[49] A. Stotland and M. Di Ventra, "Stochastic memory: enhancement due to noise", Phys. Rev. E, 85, 011116 (2012).

[50] G.A. Patterson, P.I. Fierence, and D.F. Grosz, "On the beneficial role of noise in resistive switching", Appl. Phys. Lett., 103, 074102 (2013).

[51] G.A. Patterson, P.I. Fierence, A.A. Garcia, and D.F. Grosz, "Numerical and experimental study of stochastic resistive switching", Phys. Rev. E, 87, 012128 (2013).

[52] M.S. Feali and A. Ahmadi, "Realistic Hodgkin-Huxley axons using stochastic behavior of memristors", Neural Processing Lett., 45, 1-14 (2017).

[53] S.M. Yu, "Neuro-inspired computing with emerging nonvolatile memory", Proc. IEEE, 106, 260-285 (2018).

[54] D.O. Filatov, D.V. Vrzheshch, O.V. Tabakov, A.S. Novikov, A.I. Belov, I.N. Antonov, V.V. Sharkov, M.N. Koryazhkina, A.N. Mikhailov, O.N. Gorshkov, A.A. Dubkov, A. Carollo, and B. Spagnolo, "Noise-induced resistive switching in a memristor based on ZrO2(Y)/ Ta2O5 stack", J. Stat. Mechan. Theor. Exp., 2019, 124026 (2019).

[55] S. Battistoni, R. Sajapin, V. Erokhin, A. Verna, M. Cocuzza, S.L. Marasso, and S. Iannotta, "Effects of noise sourcing on organic memristive devices", Chaos Solitons Fractals, 141, 110309 (2020).

[56] D.O. Filatov, M.N. Koryazhkina, A.S. Novikov, V.A. Shisgmakova, M.E. Shenina, I.N. Antonov, O.N. Gorshkov, N.V. Agudov, A. Carollo, D. Valenti, and B. Spagnolo, "Effect of internal noise on the relaxation time of an yttria stabilized zirconia-based memristor", Chaos Solitons Fractals, 156, 111810 (2022).

[57] O.N. Gorshkov, D.O. Filatov, M.N. Koriazhkina, V.A. Lobanova, and M.A. Riabova, "Effect of noise on resistive switching of an yttria stabilized zirconia based memristor", J. Exp. Theor. Phys., 137, 700-705 (2023).

[58] E. Salvador, R. Rodriguez, E. Miranda, J. Martin-Martinez, A. Rubio, A. Crespo-Yepes, V. Ntinas, G.C. Sirakoulis, and M. Nafria, "Noise-induced homeostasis in memristor based neuromorphic systems", IEEE Electron Device Lett., 45, 1524-1527 (2024).

[59] L.Y. Wang, J. Yang, Y. Zhu, M.D. Yi, L.H. Xie, R.L. Ju, Z.Y. Wang, L.T. Liu, T.F. Li, C.X. Zhang, Y. Chen, Y.N. Wu, and W. Huang, Rectification-regulated memristive characteristics in electron-type CuPc-based element for electrical synapse", Adv. Electronic Mater., 3, 1700063 (2017).

[60] Z.D. Liu, Y.H. Yin, X.J. Wang, S.Ju, M.Y. Song, Z. Zhou, X.J. Wang, S. Chen, Z.P. Zhang, J.Q. Liu, and W. Huang, "Two-dimensional conjugated microporous polymer with structural stability and electrical bistability for rectifying memory arrays", Adv. Intelligent Systems, 1, 1900052 (2019).

[61] L.Y. Shi, G.H. Zheng, B.B. Tian, B. Dkhil, and C.G. Duan, "Research progress on solutions to the sneak path issue in memristor crossbar arrays", Nanoscale Adv., 2, 1811-1827 (2020).

[62] S. Srivastava, J.P. Thomas, X.Y. Guan, and K.T. Leung, "Induced complementary resistive switching in forming-free $TiO_x/TiO_2/TiO_x$ memristors", ACS Appl. Mater. Interfaces, 13, 43022-43029 (2021).

[63] K.H. Kim, M.J. Seo, and B.C. Jang, "Amorphous ITZO-based selector device for memristor crossbar array", Micromachines, 14, 506 (2023).

[64] B.J. Dang, K.Q. Liu, X.L. Wu, Z. Yang, L.Y. Xu, Y.C. Yang, and R. Huang, "One-phototransistor-one-memristor array with high-linearity light-tunable weight for optic neuromorphic computing", Adv. Mater., 35, 2204844 (2023).

[65] C. Lu, J.L. Meng, J.J. Yu, J.R. Song, T.Y. Wang, H. Zhu, Q.Q. Sun, D.W. Zhang, and L. Chen, "Novel three-dimensional artificial neural network based on an eight-layer vertical memristor with an ultrahigh rectify ratio (10^7) and an ultrahigh nonlinearity (10^5) for neuromorphic computing", Nano Lett., 24, 2018-2024 (2024).

[66] S.K. Sharma, D. Haobijam, S.S. Singh, M.Z. Malik, and R.K.B. Singh, "Neuronal communication: stochastic neuron dynamics and multi-synchrony states", AEU Int. J. Electronics Communications, 100, 75-85 (2019).

[67] T. Berzina, K. Gorshkov, and V. Erokhin, "Chains of organic memristive devices: cross-talk of elements", AIP Conf. Proc., 1497, 1888-1891 (2012).

[68] A. Dimonte, T. Berzina, M. Pavesi, and V. Erokhin, "Hysteresis loop and cross-talk of organic memristive devices", Microelectronics J., 45, 1396-1400 (2014).

[69] R. Sigala, A. Smerieri, A. Schuz, P. Camorani, and V. Erokhin, "Modeling and simulating the adaptive electrical properties of stochastic polymeric 3D networks", Modelling Simul., Mater. Sci. Eng., 21, 075007 (2013).

[70] V. Erokhin, T. Berzina, K. Gorshkov, P. Camorani, A. Pucci, L. Ricci, G. Ruggeri, R. Sigala, and A. Schuz, "Stochastic hybrid 3D matrix: learning and adaptation of electrical properties", J. Mater. Chem., 22, 22881 (2012).

[71] A. Fedorova, N. Jovisic, J. Vallverdu, S. Battistoni, M. Jovicic, M. Medojevic, A. Toschev, E. Alshanskaia, M. Talanov, and V. Erokhin, "Advancing neural networks: innovations and impacts on energy consumption", Adv. Electronic Mater., in press (2024). DOI: 10.1002/aelm.202400258

[72] A. de Vries, "The growing energy footprint of artificial intelligence", Joule, 7, 1-4 (2023).

[73] A. Mikhaylov, A. Pimashkin, Y. Pigareva, S. Gerasimova, E. Gryaznov, S. Shchanikov, A. Zuev, M. Talanov, I. Lavrov, V. Demin, V. Erokhin, S. Lobov, I. Mukhina, V. Kazantsev, H. Wu, and B. Spagnolo, "Neurohybrid memristive CMOS-integrated systems for biosensors and neuroprosthetics", Frontiers Neurosci., 14, 358 (2020).

[74] S. Battistoni, M. Cocuzza, S.L. Marasso, A. Verna, and V. Erokhin, "The role of the internal capacitance in organic memristive devices for neuromorphic and sensing applications", Adv. Electronic Mater., 7, 2100494 (2021).

[75] Y. van de Burgt, E. Lubberman, E.J. Fuller, S.T. Keene, G.C. Faria, S. Agarwal, M.J. Marinella, A.A. Talin, and A. Salleo, "A non-volatile organic electrochemical device as a low-voltage artificial synapse for neuromorphic computing", Nature Mater., 16, 414-418 (2017).

Unconventional Computing in Dendrites – From the Cytoskeleton to The Spines

AVNER PRIEL

Department of Computer Science, Faculty of Engineering, Ruppin Academic Center, 4025000, Israel

Received: September 1, 2024. Accepted: September 28, 2024.

The complex electrodynamic properties of dendrites play a crucial role in information processing within neurons. This manuscript delves into the fundamental components of dendrites, focusing on the cytoskeleton and spines. By exploring the interactions between microtubules, actin filaments, ion channels, and dendritic spines, the aim is to put forward a model of unconventional computing taking place in dendrites, that contributes to neural computation and signaling. The hypothesis is that the dendritic cytoskeleton, including both microtubules and actin filaments, plays an active and central role in computations affecting neuronal function. These cytoskeletal elements are affected by, and in turn regulate, ion channels and spines activity. An integrated view of these phenomena in a bottom-up scheme is provided. We outline substantial evidence to support our model and suggest that ionic wave propagation and processing along cytoskeletal structures impact ion channels' function, and thus computational capabilities of the dendritic tree and neuronal function as a whole.

Keywords: Dendrites, cytoskeleton, microtubule, actin, unconventional computing, ionic waves, dendritic spines

1 INTRODUCTION

Neurons, the fundamental units of the nervous system, are intricate cells responsible for processing and transmitting information through complex networks. At the heart of neural communication are dendrites, specialized extensions of neurons that receive signals from neighboring cells. Under-

*Corresponding author email: avner.priel@gmail.com

standing the electrodynamic properties of dendrites is essential for unraveling the mechanisms that underlie neural signaling and computation.

The dominant view of long-term memory and information processing in biological neural networks continues to be influenced by Hebb's conjecture. The idea involves activity-depended plasticity of synaptic connections between connected neurons. In the context of neural computation (artificial neural networks), it was proved that a 3-layers network of simple, point-like neurons is capable of universal approximation. In this model, synaptic inputs are linearly added into a single threshold function.

While many models assume passive membrane properties, it is clear that the dendritic electrical properties change dynamically by modulation of voltage-gated ion channels, and further by changing its cytoskeletal components [1]. Dendritic spines provide the ability to perform nonlinear operation on post synaptic potentials locally. This means that synapses may interact in a complex way that gives rise to a set of possibilities regarding the operation and the function of dendrites. Several types of experimental evidence support this view. Neurons demonstrate significant morphological plasticity both by growing dendritic branches and by rapidly developing new dendritic spines [2]. Experiments have shown significant fluctuations of synaptic efficacy over short periods of time in response to a recent activation [3]. These findings question the model based on stable synaptic strength whose adaptation is obtained by repeated presentation of a set of training patterns.

A revised view of information processing in dendrites and nonlinear models were proposed to incorporate this evidence [4]. Among the new capabilities required from a revised model are multiplication, fast spatio-temporal correlation, and reduced sensitivity of excitatory post synaptic potential to the spatial distance from the dendrite. In this context, dynamic cytoskeletal remodeling and interactions between cytoskeletal structures and dendritic ion channel activity are required from this revised model of neuronal adaptability.

In this paper we bring to the fore the role of the cytoskeleton, particularly actin filaments, microtubules and their associated proteins, in controlling the function of dendrites. Additionally, we will discuss the electrical properties of dendritic spines, shedding light on the critical role these structures play in neural information processing.

The hypothesis is that several layers of unconventional computing contribute to the sophisticated behavior and properties of a dendrite. We shall first discuss the properties of the main constituents as necessary building blocks, and finally we shall elaborate on the suggested integrative model. In section 2, the electric properties of actin are discussed, in particular their ability to support non-linear waves. This property is essential since they serve as mediators between the entry point of electric signals in dendritic spines, to networks of microtubules in the dendritic shaft. In section 3 we further elaborate on the unique properties of microtubules as biological transistors that enable them to amplify, and manipulate signals in the form of ionic waves. These properties are supported by the ionic condensation resulted from the highly negative tubulin surface. The special properties of dendritic spines are the focus of section 4. Here we discuss the important, yet not fully understood, role of dendritic spines in mediating and modulating incoming signals from other neurons. We also discuss their interaction with actin filaments, in particular the ability of actin to modulate the spine activity. This property is essential to our model presented in section 5. In the last section we present a model integrating the above dendritic components, and suggest a possible unconventional computing mechanism taking place in dendrites. The gist of this model is as follows. Dendritic spines relay incoming signals through actin filaments to microtubule networks. These networks behave as an excitable medium and may support a form of computation akin to 'liquid state machine'. This medium property is possible due to the special organization of microtubules in dendrites in which mixed polarity elements are interspaced by microtubule associated proteins (MAP2).

2 ACTIN FILAMENTS SUPPORT NON-LINEAR IONIC WAVES

Actin is the most abundant protein in the cytoplasm of mammalian cells. Actin exists either as a globular monomer (G-actin) or as a filament (F-actin). Actin-based filaments (AF) have a diameter of approximately 7nm and variable lengths. Actin filaments are highly concentrated just below the surface of the neuronal membrane in the cell cortex. These filaments also interweave with microtubules and neurofilaments found throughout dendrites, axons, and cell bodies [5]. Spines protruding from dendrites are particularly rich in actin filaments, which frequently span from the neuronal membrane deep into the sub-spine region [6]. Actin filaments are well known to play major roles in neural growth and development, and also participate in learning related plasticity, such as the reorganization of spines following neural stimulation [7, 8]. Actin filaments also serve as anchors that stabilize the location of receptors, and ion channels. As we shall see, additional roles played by actin may derive from their nature as polyelectrolytes with charged groups that interact with counterions in the surrounding media. It follows that actin possess the ability to transmit signals via ionic conductances [9, 10]. Although there is a growing body of biophysical studies demonstrating how AF are able to act as "electric cables", this novel role proposed for AF and the underlying biophysical principles are only beginning to be understood.

Rod-like polymers, such as actin filaments, are particularly likely to have counterions adsorbed to their surface. According to the Onsager-Manning-Oosawa condensation principle, any polyelectrolyte having a charge density over 1.0 electron per Bjerrum length will have counterions adsorbed to its surface [11]. These counterions greatly increase the potential for biophysical interactions. At low ionic concentrations ions disperse in the solution surrounding actin filaments, however at high ionic concentrations, typical to the intracellular environment, ions densely attached to the surface of AF.

H. Cantiello et. al. [9] were first to report that actin filaments generate electrical signals. This conductive capability was found to depend on the counterions that adsorb to the surface of the actin filament. AF in this experiment also responded to electric fields in the range 500-2,000 V/cm. More experiments of this type revealed that actin filaments have a linear charge density of 165 e/nm enabling these filaments to act as electric charge carriers that allow ions to flow along the longitudinal axis of the filament [10]. Ions can in principle move along actin filaments without significantly affecting the surrounding environment since the ionic movement along the actin filaments is shielded. The calculated velocity of electrical signals along actin filaments is ~1–100 m/sec, which is in the range of the propagation of neural impulses, suggesting that parallel propagation along actin filaments and the axonal membrane is possible in principle.

3 MICROTUBULES PROPAGATE, AMPLIFY AND MANIPULATE ELECTRICAL SIGNALS

Microtubules (MTs) are cylindrical protein (bio)polymers of tubulin dimers, found in nearly all eukaryotic cells. They possess key roles in intracellular transport, cell organization, motion, and division. MTs are long hollow filaments made up of αβ-tubulin dimers [12] and have outer diameters measuring ~25 nm and inner diameters of ~15 nm. MTs, typically composed of 13 'protofilament' strands of αβ-tubulin, have ~1250 tubulin dimers per micron of length. Tubulin polymerization can be controlled by temperature, pH, and ion concentration to produce closely or widely spaced MTs. Importantly, MTs have an electric and functional polarity [13]. The electric properties of tubulin and MTs are of importance to a biophysical understanding of cellular processes. Theoretically MT electric properties are composed of both intrinsic protein effects and ionic effects due to condensed counter-ion clouds. These clouds arise due to the electrostatic attraction of positive ions by the negative surface charge of MTs, similarly to actin filaments. This is supported by various experimental observations [14]. Both experimental and theoretical approaches have been used to study electrical signaling along microtubules. Experimentally, the dual patch-clamp set up was used. The electrical data collected from taxol-stabilized microtubules demonstrated that microtubules behave as biomolecular transistors when responding to brief pulses of electric potential ranging ±200mV [14]. It has been found that isolated microtubules amplify the applied electric current. This suggests that microtubules exhibit a capability for ionic-based propagation that is likely to involve the condensed positive counterion cloud distributed along the length of the microtubule.

Further measurements and calculations to support the model of this phenomenon showed that microtubules support nonlinear wave propagation, much as would be expected for a biological transistor. For an electrical signal delivered to one end of a microtubule by a micropipette, the signal that arrived at the other end of the microtubule was more than twice as high as the signal in a control experiment, i.e. in solution without MTs. Conductivity along the longitudinal axis of microtubules was calculated to be on the order of 10nS, a relatively high level of ionic conductivity. Certain cations clearly participate in the positive counterion cloud that serves as a basis for this conductivity. Microtubules in solutions containing Ca^{2+} with a concentration ranging between $10^{-7} - 10^{-2}$ M increased the ability of microtubules to amplify electrical signals [14]. These data support the hypothesis that Ca^{2+} ions adsorbed to the surface of microtubules modulate the electrical properties of microtubules and further contribute to cell signaling.

The outer surface of the tubulin dimer possesses the C-termini, which extend outwardly from the MT surface and carry a significant amount of negative electric charge (as much as 40% of the total monomer charge can be attributed to a C-terminus). Since the MT surface charge is also mostly negative with positive pockets, Coulomb repulsion causes an extended C-termini conformation. Molecular dynamics modeling of the C-termini, reveals an interesting behavior [15], see Fig. 1.

The most dynamic structural elements of the MT system (i.e., its elastic and electric degrees of freedom) are envisaged as conformational states of the C-termini. Each state of the unbound C-terminus evolves so as to minimize the overall interaction energy of the system. The negatively charged C-termini interact with the dimer's surface, the neighboring C-termini and adjacent MAPs (microtubule associated protein). While the surface of the dimer is negatively charged overall, it has positive charge regions that attract the C-termini causing them to bend and bind in a 'downward' state. The energy difference between the two major metastable states is on the order of only a few k_BT [15]. Simulations demonstrate that ionic waves can trigger C-termini to change conformations from upright to downward. This model considers MAP-2 to function as an ionic "wave-guide", transferring a conformational change in a C-terminus state to an adjacent MT. Perturbations applied to condensed counterions at the end of MAP-2 drive them from equilibrium, initiating a travelling wave. This wave is predicted to behave as a "kink" solitary wave with a (high) phase velocity of about 2nm/ps (2km/s).

Interestingly, the structure of the MT is more complex, being cylindrical in shape with a hollow interior and an additional negative linear charge density along its inner surface. The lattice-like wall interspersed with nanopores having an effective radius of 4-5Å, connecting the inner and outer surfaces, reminiscent of ion channels. This radius is enough to be permeable to the counterions covering both sides of the MT. Based on the results of dynamic simulations [16], the nanopores were found to have asymmetric inner and outer conductances. A combination of this asymmetry and an internal voltage source arising from the motion of the C-terminal tails, or electric potential perturbation of external source, causes a net current to be pumped across the microtubule wall and propagate down the microtubule through the lumen. This effect enhances and adds directly to the longitudinal current through the lumen, and could be significant in amplifying weak endogenous currents within the cellular environment.

Although the model described in the last section is classical, it is worth mentioning that several papers have discussed the possibility of quantum properties of tubulin dimers and their implications.

4 ELECTRICAL PROPERTIES OF DENDRITIC SPINES ARE MODULATED BY ACTIN

Dendritic spines (DS) are small protrusions on dendrites that play a crucial role in excitatory postsynaptic responses. These structures are the focus of a high density of ion channels and dynamic actin networks, creating specialized compartments for signal processing. DS are highly dynamic structures and are characterized by their morphological diversity thought to provide a substrate for memory storage and synaptic transmission [17,18]. In the hippocampus, spines show large variation in size and shape [19]. DS appear in different shapes and forms based on the relative sizes of the spine head and neck. Among the prevalent shapes, we mention the mushroom spines with a large head and a narrow neck, thin spines with a smaller head and a narrow neck, and stubby spines [20]. Although voltage-sensitive dye labeling revealed that the DS may not be isolated (electrically) from the dendrite [21], the electrical function of the DS has largely remained an open question [22]. Direct electrical measurements of ionic conductance in DS remains largely lacking, mainly due to their size and dynamic behavior. Nevertheless, several optical methods have been successfully applied [23, 24]. The functional role of microtubules in DS is also not clear, although it was found that actin and microtubule crosstalk in controlling synapse function and plasticity, and that structural changes in the actin cytoskeleton at the base of the spine are directly involved in microtubule invasion into DS [25,26].

In [27], several experimental techniques have been used to obtain direct electrical information from DS of adult mouse hippocampus. DSs were isolated from dissected hippocampus. Lipid bilayer membrane (BLM) reconstitution and patch clamping techniques were employed to obtain electrical signals from ion channels present in the DS membrane. Additionally, patch clamped DS in cultured mouse hippocampal neurons was used to validate the electrical information observed with the isolated spines. The main results of this study are given below. Single-channel currents in the reconstituted DS were observed. The channels had a single channel conductance of ~50pS and were further blocked by the addition of Zn^{2+}. Furthermore, spontaneous activity in the form of electrical oscillations was observed after reconstitution of DS membranes in the BLM reconstitution system. This phenomenon was further explored with membrane-attached patch clamping both in vitro and in situ. The oscillations observed suggested that DS membranes contain the required active electrical components to support spontaneous generation of electrical signals. Interestingly, the frequency of this signal was strongly dependent on the holding potential, with almost complete inhibition at –20mV. Addition of Ca^{2+} after the generation of the oscillations, caused a dramatic shift in the frequency and "reversal potential" of the spontaneous activity. This finding suggests the presence of Ca^{2+}-dependent oscillatory behavior, which is intrinsic to DS function, and may hint on the possible role of spines as dynamic (tunable) filters.

Another interesting result was observed in the in-situ experiments where electrical signals generated by the DS back propagated along dendrites. Using double-electrode patch clamping, oscillations were found to be coupled between DS and dendrites, suggesting the signal on dendrites are propagated from DS. There were usually very short delays (~20 ms) between the DS-generated waveforms, and those evoked at the dendritic site.

Finally, it was found that the actin cytoskeleton regulates ion channel activity in DS. Addition of cytochalasin D (actin filament disrupter) increased instances of channel activation. In the absence of NMDA, addition of actin and ATP was sufficient to activate channel currents. The Actin cytoskeleton regulates DS ion channel activity, suggesting that dynamic changes in cytoskeletal organization help modulating DS electrical activity. Since Actin filaments and MTs interact with each other in the formation of dendritic networks, it is expected that cytoskeletal biopolymers may contribute to the intrinsic electrical properties of DS.

This data suggest that DS electrical activity is a complex interplay between membrane ion channels and the underlying cytoskeleton. Thus, the process of cytoskeleton remodeling feeds back on DS ion channels. It further demonstrates that DS are active sites of ion channel cytoskeleton interactions, hence it demonstrates and supports the idea that DS are highly elaborate electrical compartments that contribute actively to neuronal behavior.

5 UNCONVENTIONAL COMPUTING MODEL IN DENDRITES

Classical ('conventional') models of computation are based on the concept of Turing machines. Consequently, every task the computing machine is required to process has to be pre-programmed (algorithm). The computational process runs until a halt-state is invoked. In this framework the input is introduced at the beginning of the operation; however, no trace/memory of this computation remains when the same machine carries out a subsequent computation. One of the paradigms proposed to overcome the apparent discrepancy in solving complex, as well as temporal problems were artificial neural networks. These computational models rely on massively interconnected parallel networks of simple units, as well as on the availability of some learning algorithm to train the parameters of the model given representative data regarding the problem. Artificial neural network models have been widely used, and in the past few years gained considerable boost in the scale and variety of applications and problems tackled. A lot has been argued for and against the capabilities and bounds of neural networks, however it is acceptable that even very large and deep networks can be viewed as a high-dimensional complex probability distribution of its input/output space. While synapses (weights) in artificial neural networks vary slowly during learning, or are taken to be static after the learning phase is over, real synapses in neurons are highly dynamic and activity-dependent. This means that the neuron itself is much more than a simple processing unit, unlike its artificial model.

In view of the extensive body of research and accumulated data that support the idea of dendritic computation, an alternative hypothesis has been proposed that relies on unconventional computing taking place in dendrites. There are several mechanisms that can contribute to this alternative, spanning several scales and levels. Starting from the smallest scale, the C-termini on MTs interact with their neighboring C-termini, as well as with MAPs, as described above. This nonlinear interaction controls the propagation of signals along MTs, and between MTs through interconnecting MAPs. Although this mechanism may seem passive, it is in fact dynamic since the C-termini are sensitive to electric fields as well as environmental factors, e.g. pH. This sensitivity is translated to different patterns of wave propagation along the MTs. A schematic diagram of this scenario is depicted below. In figure 1, a

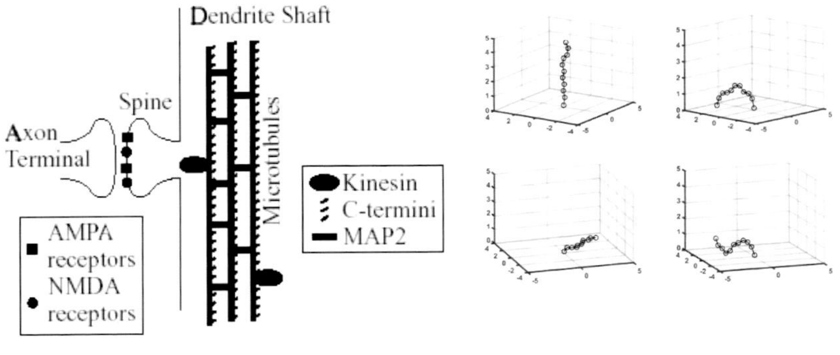

FIGURE 1
Left - C-termini on MTs within the dendritic shaft. Right - possible conformations of C-termini obtained from molecular dynamics simulation (C-termini length ~4nm) (images from [28,15])

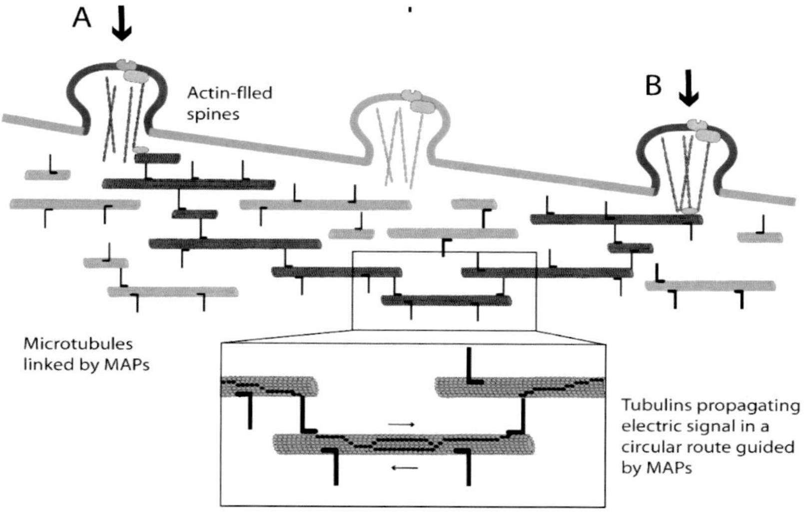

FIGURE 2
Electric signals coincidence from two ports (spines). The insert depicts a closeup on a possible wave pattern. (image from [29])

small segment of the dendritic shaft is shown with a network of MTs decorated by c-termini, interspaced by MAP2 which are highly abundant in the dendritic region of the neuron. Adjacent dendritic spine is also shown which represents an electric terminal to the subsequent circuit.

Figure 2 shows a larger segment of a dendrite where input signals from two spines may propagate through actin in the spines to MT networks. These coincident inputs are spatiotemporal, i.e. both space and time dependent, as the time difference between the incoming signals may vary. It is also possible that the signals are in fact pulse-trains, therefore the wave pattern in the subcircuit shown in the insert would be more complex.

The next level/layer of the model comprised of voltage sensitive ion channels (in dendritic membrane and spines) as possible output terminals. Actin networks interact and control ion-channels activity through several regulatory mechanisms [27, 31]. As shown schematically in figure 3, synaptic input signals propagate via actin bundles within the synaptic neck to the dendritic shaft, or directly to MT networks (MTN). The MTN operates as a high dimensional dynamic system, which continuously processes the incoming perturbation. Within the MTN one can expect recurrent operations due to MAP2 coupling between adjacent MTs (curved arrows in the figure). The temporal state field of the MTN (can be viewed as an analogue state machine) is used as 'readout' at certain areas of the dendritic shaft, by interacting with membrane ion channels. This view links dendritic components, (namely

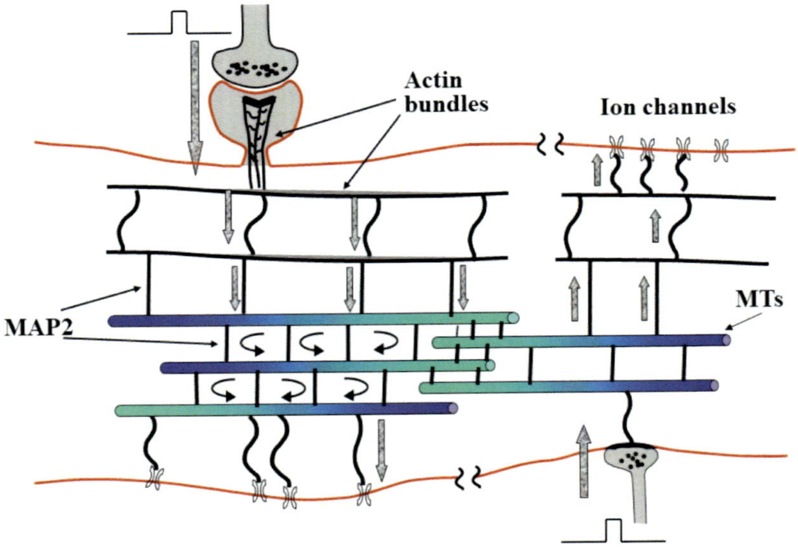

FIGURE 3
Schematic picture of the processing model (image adapted from [30])

spines, ion channels, actin bundles within the spine and along the membrane, and MTNs) to their position in the dendritic shaft.

Figure 4 shows a functional diagram of the same model. This figure highlights the signal processing aspects of the model. The model proposes that a MTN is capable of generating diverse electric patterns (or more generally, phase space trajectories) in response to different input vectors. From the modeling point of view, the requirements from such a system are not too restrictive since the output-state is not an attractor of the system. In other words, information processing at this level is not necessarily based on attractor dynamics but rather on real time computations. This means that the output is being "read" continuously and not at certain times (clock), or when the system relaxes to a final state. This proposition

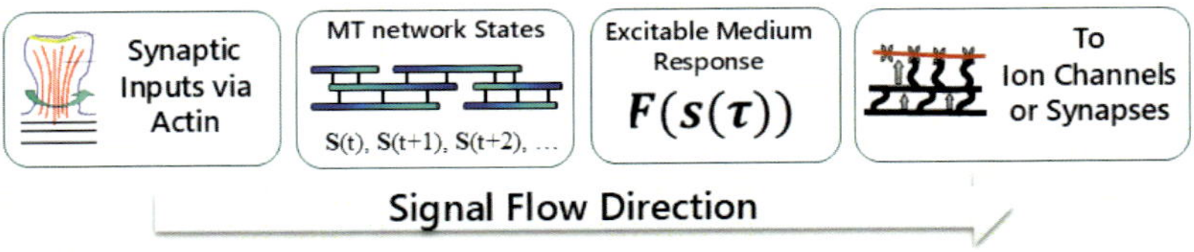

FIGURE 4
Functional diagram of the signal flow and processing. Synaptic inputs are transferred to the Microtubule network via Actin filaments, perturbing the temporal state field. The network's state at time 'τ' is transformed by the function F and further transmitted internally or to target outputs, i.e. ion-channels or dendritic spines.

is based on the previously discussed observations about the MTN's ability to propagate signals, as well as the unique spatial topology of MTNs in dendrites, where mixed-polarity MTs are interconnected by MAP2s.

The output from the MTN would be some function of the evolved state vector in certain areas accessed by actin filaments and/or directly linked to ion channels (denoted by $F(s(\tau))$ in the third block of figure 4). These output signals may modify the temporal channel activity, either by directly arriving from the MTN to the channels, or mediated by actin filaments. Hence, the channel-based synaptic membrane conductance is regulated.

The idea that a non-specific high dimensional dynamical system may serve as a reservoir of trajectories in the context of liquid state machines (LSM), has been suggested to explain the role of microcircuits in the brain, e.g. [32]. The basic structure of an LSM is an excitable medium and an output function that maps the current liquid state. The medium must be sufficiently complex and dynamic to guarantee a universal computational power. This may be sufficient to ensure that different input vectors will lead to separate trajectories. A network of spiking neurons, a recurrent neural network, and even real water have been used as a `liquid', whereas the output (readout) function has been implemented by either simple perceptrons, threshold functions or even linear regression functions. The analogy of our model depicted in figures 3,4 is clear, where the microcircuits of our model are the microtubule-networks. It is worth mentioning that there can be plenty of MTN circuits in dendrites since in a cross section of a typical dendrite of diameter 1-1.5mm, there can be several dozens of microtubules. In view of their alternating polarity order and interconnections via MAP2s, one can deduce that many circuits can co-exist.

6 DISCUSSION

Recent findings have highlighted the limitations of the prevailing models, particularly in accounting for the dynamic and nonlinear properties of dendrites. In this manuscript we propose an alternative view that incorporates the complex interactions between dendritic components and their biophysical properties. These properties enable dendrites to perform nonlinear operations on postsynaptic potentials, thereby allowing for a more complex form of signal processing.

The involvement of cytoskeletal components, particularly actin filaments and microtubules, is crucial in this revised model of dendritic computation. Actin filaments, for example, have been shown to support nonlinear ionic wave propagation, a property that is essential for their role as mediators between the entry point of electric signals in dendritic spines and the networks of microtubules in the dendritic shaft. This ability to transmit signals via ionic conductances suggests that actin filaments may function as "electric cables," a concept that is only beginning to be understood but holds significant implications for our understanding of neural processing. Microtubules have been proposed to function as biological transistors, capable of amplifying and manipulating electrical signals. The experimental evidence supporting this view is growing, with studies demonstrating that microtubules respond to electric and magnetic fields, and support nonlinear wave propagation as well. In the realm of the dendritic shaft, the combination of microtubules and their associated proteins, mainly MAP2, generate microcircuits composed of MTs of mixed polarity. This interesting property can be responsible for recurrent waves that interact nonlinearly.

Dendritic spines, being rich in both actin filaments and ion channels, are also highlighted as critical components of this revised model. The spines are not only sites of synaptic input but also play an active role in signal processing, with their electrical properties modulated by the underlying actin cytoskeleton. This modulation suggests a feedback loop in which cytoskeletal remodeling influences spine activity, which in turn affects the overall computational capacity of the neuron.

The proposed unconventional computing model for dendrites, goes beyond the classical Turing machine framework. This model postulates that dendritic microtubule networks (MTNs) function as high-dimensional dynamic systems capable of real-time computation. The interactions between MTNs, actin filaments, and ion channels create a complex, nonlinear system that can process spatiotemporal inputs in a manner akin to a "liquid state machine". This model may provide a framework for understanding how these properties contribute to the sophisticated computational abilities of neurons.

In summary, the alternative view of dendritic function presented here represents a significant departure from traditional models of neural computation. By incorporating the dynamic properties of cytoskeletal components and their interactions with ion channels and dendritic spines, this model may offer a more comprehensive understanding of how neurons process information. This paradigm shift opens new avenues for exploring the computational capabilities of neurons and their role in higher-order brain functions.

REFERENCES

[1] Kaech, S., H. Parmar, M. Roelandse, C. Bornmann, and A. Matus. 2001. Cytoskeletal microdifferentiation: A mechanism for organizing morphological plasticity in dendrites. Proc. Natl. Acad. Sci. USA. 98:7086–7092.

[2] Hering H, Sheng M. Dendritic spines: structure, dynamics and regulation. Nat Rev Neurosci. 2001 Dec;2(12):880-8.

[3] Hempel, C. M., K. H. Hartman, X. J. Wang, G. G. Turrigiano, and S. B. Nelson. 2000. Multiple forms of short-term plasticity at excitatory synapses in rat medial prefrontal cortex. J. Neurophysiol. 83:3031-3041

[4] Righes Marafiga J, Calcagnotto ME. Electrophysiology of Dendritic Spines: Information Processing, Dynamic Compartmentalization, and Synaptic Plasticity. Adv Neurobiol. 2023;34:103-141. doi: 10.1007/978-3-031-36159-3_3. PMID: 37962795.

[5] Goldman JE. Immunocytochemical studies of actin localization in the central nervous system. J Neurosci. 1983 Oct;3(10):1952-62.

[6] Honkura N, Matsuzaki M, Noguchi J, Ellis-Davies GC, Kasai H. The subspine organization of actin fibers regulates the structure and plasticity of dendritic spines. Neuron. 2008 Mar 13;57(5):719-29.

[7] Star EN, Kwiatkowski DJ, Murthy VN. Rapid turnover of actin in dendritic spines and its regulation by activity. Nat Neurosci. 2002 Mar;5(3):239-46.

[8] Matus A. Actin-based plasticity in dendritic spines. Science. 2000 Oct 27;290(5492):754-8.

[9] Cantiello HF, Patenaude C, Zaner K. Osmotically induced electrical signals from actin filaments. Biophys J. 1991 Jun;59(6):1284-9.

[10] Tuszynski JA, Portet S, Dixon JM, Luxford C, Cantiello HF. Ionic wave propagation along actin filaments. Biophys J. 2004 Apr;86(4):1890-903.

[11] Trizac E, Tellez G. Onsager-Manning-Oosawa condensation phenomenon and the effect of salt. Phys Rev Lett. 2006 Jan 27;96(3):038302.

[12] Dustin, P. 1984. Microtubules. Springer-Verlag.

[13] Tuszynski, J. A., Craddock, T. J. A. & Carpenter, E. J. 2008. Bio-Ferroelectricity at the Nanoscale. Journal of Computational and Theoretical Nanoscience 5: 2022-2032.

[14] Priel, A., Ramos, A. J., Tuszynski, J. A. & Cantiello, H. F. 2006. A biopolymer transistor: electrical amplification by microtubules. Biophysical Journal 90: 4639-43.

[15] Priel, A., Tuszynski, J. A. & Woolf, N. J. 2005. Transitions in microtubule C-termini conformations as a possible dendritic signaling phenomenon. European Biophysics Journal 35: 40-52.

[16] Freedman, H., Rezania, V., Priel, A., Carpenter, E., Noskov, S.Y., and Tuszynski, J.A., Model of ionic currents through microtubule nanopores and the lumen. Phys. Rev. E. 2010. May 81(5):051912

[17] Mahmmoud RR, Sase S, Aher YD, Sase A, Gröger M, Mokhtar M, Höger H, Lubec G. Spatial and Working Memory Is Linked to Spine Density and Mushroom Spines. PLoS One. 2015 Oct 15;10(10):e0139739.

[18] Leuner B, Shors TJ. New spines, new memories. Mol Neurobiol. 2004 Apr;29(2):117-30

[19] von Bohlen Und Halbach, O. (2009). Structure and function of dendritic spines within the hippocampus. Ann. Anat. 191, 518–531

[20] Tønnesen J, Nägerl UV. Dendritic Spines as Tunable Regulators of Synaptic Signals. Front Psychiatry. 2016 Jun 9;7:101

[21] Popovic, M., Carnevale, N., Rozsa, B. et al. Electrical behaviour of dendritic spines as revealed by voltage imaging. Nat Commun 6, 8436 (2015)

[22] Zecevic D. Electrical properties of dendritic spines. Biophys J. 2023 Nov 21;122(22):4303-4315

[23] Araya R, Vogels TP, Yuste R. Activity-dependent dendritic spine neck changes are correlated with synaptic strength. Proc Natl Acad Sci U S A. 2014 Jul 15;111(28):E2895-904.

[24] Jayant K, Hirtz JJ, Plante IJ, Tsai DM, De Boer WD, Semonche A, Peterka DS, Owen JS, Sahin O, Shepard KL, Yuste R. Targeted intracellular voltage recordings from dendritic spines using quantum-dot-coated nanopipettes. Nat Nanotechnol. 2017 May;12(4):335-342.

[25] Schätzle, P., Esteves da Silva M., Tas R.P., Katrukha E.A., Hu H.Y., Wierenga C.J., Kapitein L.C., and Hoogenraad C.C.. 2018. Activity-dependent actin remodeling at the base of dendritic spines promotes microtubule entry. Curr. Biol. 28:2081–2093.e6. 10.1016

[26] Peña-Ortega F, Robles-Gómez ÁA, Xolalpa-Cueva L. Microtubules as Regulators of Neural Network Shape and Function: Focus on Excitability, Plasticity and Memory. Cells. 2022 Mar 8;11(6):923

[27] Priel A, Dai X-Q, Chen X-Z, Scarinci N, Cantero MdR and Cantiello HF. Electrical recordings from dendritic spines of adult mouse hippocampus and effect of the actin cytoskeleton. Front. Mol. Neurosci. 2022 15:769725

[28] Priel, A., Tuszynski, J.A., Cantiello, H.F. (2006). The Dendritic Cytoskeleton as a Computational Device: An Hypothesis. In: Tuszynski, J.A. (eds) The Emerging Physics of Consciousness. The Frontiers Collection. Springer, Berlin, Heidelberg.

[29] Woolf J Nancy, Priel Avner, Tuszynski A Jack. (2010). Nanoneuroscience: structural and functional roles of the neuronal cytoskeleton in health and disease, Springer-Verlag Berlin Heidelberg

[30] Priel, A., Tuszynski, J.A. & Woolf, N.J. Neural cytoskeleton capabilities for learning and memory. (2010) J Biol Phys 36, 3–21

[31] Shaw, J.E., Koleske, A.J. (2021) Functional interactions of ion channels with the actin cytoskeleton: does coupling to dynamic actin regulate NMDA receptors? The Journal of Physiology 599(2): 431-441

[32] Maass W, Natschlager T, Markram H. (2002) Real-time computing without stable states: a new framework for neural computation based on perturbations. Neural Comput. Nov;14(11):2531-60

Interaction of Chemical Reaction and its Environment Enhancing the Advantage of Unconventional Computation

YUKIO PEGIO GUNJI*

Department of Intermedia Art and Science, School of Fundamental Science and Technology, Waseda University, Ohkubo 3-4-1, Shinjuku-ku, Tokyo, 169-8555, Japan

Received: September 1, 2024. Accepted: September 15, 2024.

We implemented a chemically inspired cellular automaton using a model of an artificial chemical system as a molecular ensemble, in which the degree of self-aggregation and self-dispersion interacts with the degree of activation. This molecular ensemble is designed to periodically repeat self-aggregation (clustering) and dispersion (de-clustering) in the absence of noise. However, it was shown to exhibit stationary fluctuations under noise conditions, particularly 1/f noise, which is characteristic of critical phenomena. Critical phenomena are suggested to have high computational power. Here, the trade-off between computational universality and computational efficiency was evaluated. As a result, it was shown that while ordinary cellular automata display a clear trade-off between computational universality and computational efficiency as defined here, the chemically inspired cellular automata break this trade-off. This is one of the powerful advantages of unconventional computing.

Keywords: Chemical computing, unconventional computing, trade-off principle, cellular automata, critical phenomena

1 INTRODUCTION

Chemical reactions, especially biochemical reactions in cells, do not occur in well-stirred solutions as described in traditional chemical reaction theory.

* Contact author: E-mail: pegioyukio@gmail.com

Instead, they take place in heterogeneous, extremely complex reaction systems where molecules repeatedly undergo self-aggregation and self-dispersion throughout the system [1–8]. Moreover, the dynamics of these molecular aggregates alter the microscopic boundary conditions of the chemical reactions, making it exceedingly difficult to understand the reaction system using a dynamical systems approach with the solution concept as a variable. To address such a reaction system, it is necessary to consider the chemical reactions at the molecular level. Chemical reaction models using Gillespie's algorithm [9–11] and multi-sets [12, 13] are particle-based reaction models. However, these models do not account for molecular structures, self-aggregation, and self-dispersion, nor do they consider structure-dependent changes in reaction constants.

We have recently proposed a chemical reaction system that repeats self-aggregation and self-dispersion, with reaction constants that change depending on the size of the cluster. We outlined this reaction and demonstrated its potential for implementing logic gates [14]. In this paper, we define this system in more detail, investigate its behavior, and distribute the molecular population into multiple cells (chambers). These cells interact using an enzyme, enabling the implementation of a cellular automaton [15–17] inspired by chemical reactions. Specifically, we show that adding fluctuations to chemical reactions dependent on the state created by molecular population-sachieved through self-aggregation and self-dispersionresults in power-law distributions and 1/f noise, which are characteristic of self-organized criticality [18–21] and phenomena known as the edge of chaos [22–24].

Recently, advances in investigating the firing states of neurons in the brain have revealed many power-law distributions related to the avalanche phenomenon of firing and the time distribution of firing [25–27]. These findings are thought to be related to self-organized criticality and/or the edge of chaos, suggesting high computational power [28–30]. Indeed, the rules for implementing the universal Turing machine in elementary cellular automata (ECA) are composed of Rule 110 (class IV) [31], which is thought to be at the edge of chaos between automata showing order (classes I and II) and those exhibiting chaos (class III) [17]. However, the prevailing interpretation is that the edge of chaos and critical phenomena are extremely rare and that these rare phenomena have been selected by natural selection.

However, the authors believe that viewing the edge of chaos as a rare phenomenon at a phase transition point is an extremely limited perspective when considered in the context of natural phenomena. Cellular automata are typically constrained to synchronous updating, and the state concept in dynamical systems often assumes, for example, a homogeneous animal population in an ecosystem. When cellular automata are redefined based on asynchronous updating [32–34], it has been shown that there are rather few rules that indicate chaos or order, with most exhibiting criticality [35–37]. This is referred to as unfolded criticality. In other words, natural phenomena are far from the ideal situations assumed by conventional theories and are much more complex. From this perspective, critical phenomena are not rare.

The chemical reaction system discussed in this paper, which repeats self-aggregation and self-dispersion and changes reaction constants depending on the size of the cluster, fundamentally differs from such a dynamical system model that assumes the usual concept of a solution. Therefore, its behavior is likely closer to unfolded criticality.

In this paper, we place this chemical reaction system in multiple independent cells, interact with it using an enzyme, and implement a cellular automaton inspired by chemical reactions. We discuss its computational capabilities, particularly its computational universality and computational efficiency [38]. Typically, computational universality is discussed from the perspective of Turing machines. However, here we define universality as the degree to which the system can reach a number of possible states through computation, and computational efficiency as how quickly it can reach a reachable state. From this perspective, we compare how chemically inspired cellular automata differ from ordinary automata.

2 CHEMICAL REACTION IN CLUSTERING AND DE-CLUSTERING

We previously proposed a chemical reaction system in which molecules repeatedly undergo self-aggregation and self-dispersion, with the reaction constant depending on the size of the molecular clusters. Here, we define that chemical reaction system more mathematically.

The N molecules are initially prepared under a specific perturbation of p_{noise}, where the rate of activated molecules are randomly given. The time step of the chemical reaction is represented by τ. The partition of molecules of at the τth step is expressed as

$$P^{\tau} = \{\{m^{\tau}(a), ..., m^{\tau}(b)\}, \{m_i^{\tau}(c), ..., m^{\tau}(d)\}, ..., \{m^{\tau}(e), ..., m^{\tau}(f)\}\}, \quad (1)$$

where $a, b, ..., f \in \{1, ..., N\}$ and $a \neq b \neq ... \neq f$. $M = \{m^{\tau}(a), ..., m^{\tau}(f)\}$, represents a collection of molecules at the τth step. Elements of a partition represent clusters and are designated as follows.

$$C^{\tau}(1) = \{m^{\tau}(a), ..., m^{\tau}(b)\}, \quad (2)$$

$$C^{\tau}(2) = \{m^{\tau}(c), ..., m^{\tau}(d)\}, ..., \quad (3)$$

$$C^{\tau}(s^{\tau}) = \{m^{\tau}(e), ..., m^{\tau}(f)\}. \quad (4)$$

The value of s^τ represents the number of clusters at the τth step, such that

$$s^\tau = \#P^\tau \tag{5}$$

where $\#S$ of a set S represents the number of elements of S. Adapting $v : M \to \mathbf{B}$ to $m^\tau(k)$, the state of molecule which is either active (1) or inactive (0), such as $v(m^\tau(k)) = 0$ or 1, where $\mathbf{B}=\{0, 1\}$. Thus the number of active molecule at the τth step is expressed as

$$a^\tau = \#\{m^\tau(k) \in M | v(m^\tau(k)) = 1\}. \tag{6}$$

Chemical reaction proceeds by receiving the noise, clustering, de-clustering, activation and inactivation. At first, the state of molecules is flipped with probability, p_f. Receiving the noise is accomplished under p_f as

$$v(m^\tau(k)) = (v(m^\tau(k)) + 1) mod 2. \tag{7}$$

The reason why the time step τ at the left hand is the same as that at the right hand in the equation is what the state change proceeds asynchronously.

The clustering is accomplished by the following. Two clusters (i.e., elements of partition), $C^\tau(k)$ with $k = w$ or $k = q$ and $w \leq q$ and $w \neq q$ are randomly chosen and if the clusters satisfy the condition such that

$$\#\{m^\tau(k) \in C^\tau(k) | v(m^\tau(k)) = 1\} / \#C^\tau(k) \leq p_{cl} \tag{8}$$

and $s^\tau \neq 1$ then the clusters are united into one cluster by

$$C^\tau(w) = C^\tau(w) \cup C^\tau(q). \tag{9}$$

such that for left-hand term,

$$C^\tau(w) = \{m^\tau(w_1), ..., m^\tau(w_m), m^\tau(q_1), ..., m^\tau(q_n)\} \tag{10}$$

where for right-hand term,

$$C^\tau(w) = \{m^\tau(w_1), ..., m^\tau(w_m)\}, C^\tau(q) = \{m^\tau(q_1), ..., m^\tau(q_n)\}. \tag{11}$$

As the allocation of $C^\tau(q)$ becomes vacant, the allocation for q-cluster will be empty, and the indexes of the clusters after q will be shifted forward such that

$$C^\tau(x) = C^\tau(x + 1) \tag{12}$$

for $x \geq q$.

The de-clustering is defined as the following. One cluster (i.e., an element of partition), $C^\tau(h)$ is randomly chosen and if the cluster satisfies the condition such that

$$\#\{m^\tau(k) \in C^\tau(k) | v(m^\tau(k)) = 1\} / \#C^\tau(k) \geq 1 - p_{cl} \tag{13}$$

and $s^\tau \neq \#M$ (i.e., all molecules are not monomers) then the cluster is divide into two clusters such that after de-clustering,

$$C^\tau(h) = \{m^\tau(h_1), ..., m^\tau(h_m)\}, C^\tau(s^\tau) = \{m^\tau(h_{m+1}), ..., m^\tau(h_n)\}. \tag{14}$$

where before de-clustering

$$C^\tau(h) = \{m^\tau(h_1), ..., m^\tau(h_m), m^\tau(h_{m+1}), ..., m^\tau(h_n)\}, \tag{15}$$

and the value of h_m is randomly determined. It implies that newly constructed cluster is allocated for the end of the line.

Activation of the molecules is defined by the following. If $s^\tau = 1$ (i.e., all molecules constitute one giant cluster) for any $k \in \{1, 2, ..., \#M\}$

$$v(m^\tau(k)) = 1. \tag{16}$$

Inactivation is similarly defined by the following. If $s_i^\tau = \#M$ (i.e., all molecules are monomers) for any $k \in \{1, 2, ..., \#M\}$

$$v(m^\tau(k)) = 0. \tag{17}$$

It is clear to see that once all molecules constitute a giant cluster, the situation of which all molecules are activated as far as de-clustering continues, and that once all molecules become a monomers, the situation of which all molecules are inactivated as far as clustering continues. It results in stable square waves if there is no noise. After clustering, de-clustering, activation and inactivation, all states are updated for any $k \in \{1, 2, ..., \#M\}$

$$m^{\tau+1}(k) = m^\tau(k), \tag{18}$$

and for any $j \in \{1, 2, ..., s^\tau\}$

$$P^{\tau+1} = P^\tau. \tag{19}$$

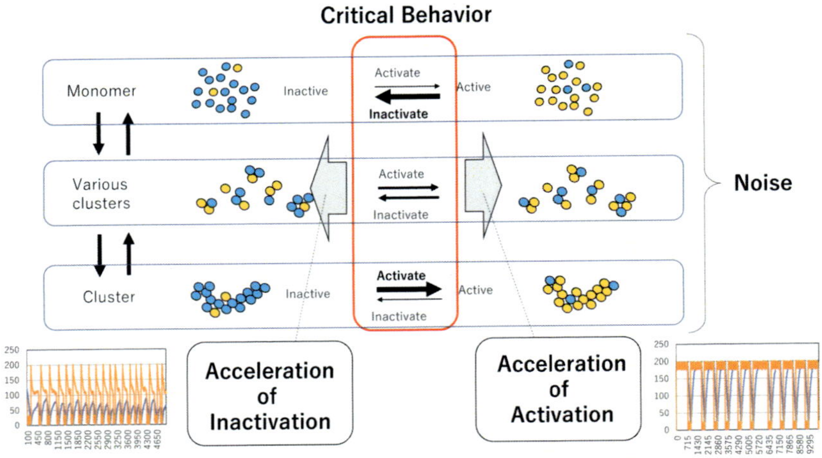

FIGURE 1
Schematic diagram of the chemical reaction interacting the size of clusters of molecules. Blue and yellow circles represent inactive and active molecules, respectively. The horizontal axis represents the axis of activation-inactivation of molecules, and the vertical axis represents the size of clusters. The range surrounded by red line represents fluctuated states of molecules influenced by the noise. The effect of the enzyme is represented by big gray arrows.

By these updates, both states do not changed and changed are updated. Depending on these updating, the number of activated molecules and the number of clusters are also updated.

In addition, we introduce the administration of an enzyme. If the proportion of inactive molecules among all molecules is between 0.3 and 0.7, and the number of inactive molecules is greater than that of active molecules, the administration of the enzyme for inactivation makes the $v(m^{\tau}(k))$ of all molecules 0 with a probability of p_e=0.95. Similarly, if the proportion of inactive molecules among all molecules is between 0.3 and 0.7, and the number of active molecules is greater than that of inactive molecules, the administration of the enzyme for activation makes the $v(m^{\tau}(k))$ of all molecules 1 with a probability of p_e=0.95.

Figure 1 shows a schematic diagram of the chemical reaction interacting with the size of clusters of molecules. It illustrates how the chemical reaction proceeds with the reaction constant depending on the cluster size, changing the cluster size through self-aggregation and self-dispersion. There are both directions of activation and inactivation over all cluster sizes under noise conditions, while monomers generally become inactive and molecules in the largest cluster become active.

The region surrounded by the red loop represents the population of molecules under noisy conditions. As mentioned later, the chemical

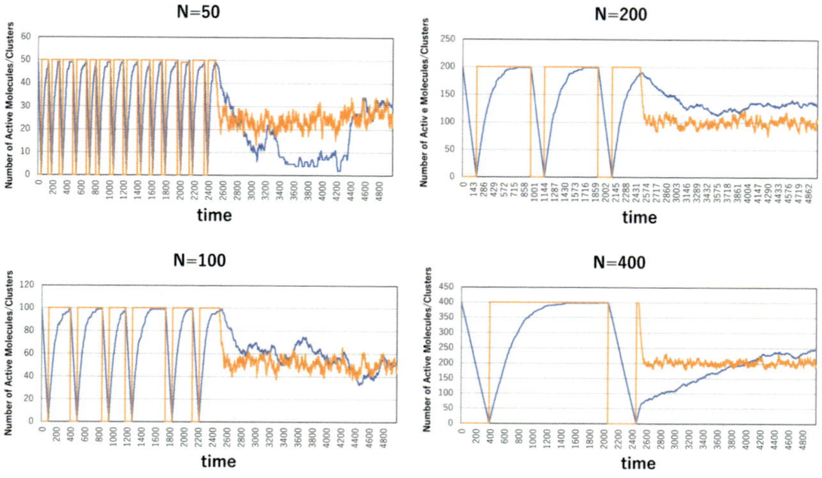

FIGURE 2
The number of active molecules (orange) and clusters (blue) against time. Each graph represents the time series of different number of molecules indicated by N. The chemical reaction proceeds without noise from the initial to the 2500 time steps with P_{cl}=0.4, and it proceeds with noise with P_f=0.02 after 2500 time steps.

molecules lead to cyclic oscillation without noise, and they lead to a fluctuating steady state under noisy conditions. The large gray arrows in Figure 1 show that applying the enzyme to the chemical molecules under noisy conditions accelerates both inactivation and activation. The accompanying graphs show the time series of the number of clusters and active molecules after the administration of enzymes.

Figure 2 shows four cases (N=50, 100, 200, and 400) of a time series of the number of activated molecules and clusters. Initially, all molecules are inactive monomers, and this is under no noise conditions before 2500 steps. It is clear that cyclic oscillation is iterated until the noise is introduced. Clustering continues gradually until all molecules constitute the largest cluster. Once all molecules are recruited into one largest cluster, all molecules are activated, and de-clustering begins and continues until all molecules become monomers. Soon after all molecules become monomers, all molecules become inactivated, and then clustering begins. This process is repeated until noise is introduced. Since the cycle period is determined by the time needed to constitute the largest cluster consisting of (N molecules, the larger the system size, the longer the period.

Figure 2 also shows that noise with p_f=0.02 causes the population of molecules to fluctuate such that the number of activated molecules is maintained equal to the number of inactivated molecules, and the number of clusters fluctuates across various sizes.

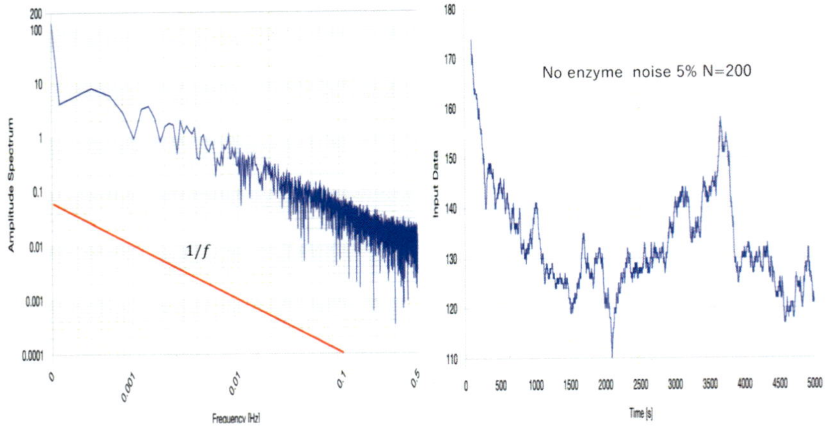

FIGURE 3
The power spectrum against frequency (left) for a time series of the number of clusters with N=200 (right). The noise condition is P_f=0.05

How is the fluctuation under noisy conditions characterized? Figure 3 shows the characterization of the fluctuated time series of the number of activated molecules under noisy conditions with p_f=0.05. It is typical $1/f$ noise, which is expressed as a linear relation of the power spectrum against frequency in a log-log plot. Notably, the exponent of the line is almost exactly -1.0. We estimated various noise levels for the population of molecules with N=200, such as p_f=0.01 0.05, and all time series exhibit exact $1/f$ noise independent of the noise size.

This kind of power law is characteristic of critical phenomena. The idea of self-organized criticality and/or the edge of chaos posits that a biological system is autonomously tuned to the critical state at the phase transition between chaos and order. It is also claimed that a system at the edge of chaos can achieve high computational performance, transcending the trade-off principle between computational universality and highly efficient computation.

Figures 4 and 5 show the effects of the enzyme in accelerating the inactivation and activation of molecules, respectively. Each graph represents a time series of the number of clusters (blue line) and the number of active molecules (orange line) under various conditions. A pair of graphs surrounded by a red loop represents no noise condition (upper pair) and noise condition with P_f=0.02 (lower pair). A pair of graphs surrounded by a blue loop represents the administration of the enzyme condition (left pair) and no enzyme condition (right pair).

From the graphs in Figure 4, it can be observed that the enzyme accelerating the inactivation of molecules contributes to the generation of spike trains

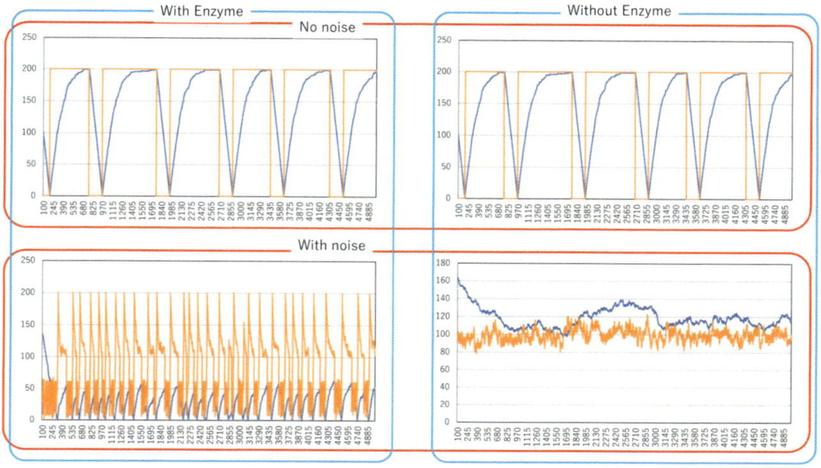

FIGURE 4
The effect of the enzyme to accelerate inactivation of molecules. Blue lines represent a time series of the number of clusters, and orange lines represents a time series of the number of active molecules.

under noise conditions, but not under no-noise conditions. Since the enzyme is administered when inactive molecules are dominant (i.e., the rate of inactive molecules is 0.50.7) in the population of molecules, the number of active molecules is perpetually maintained at a smaller level, where it fluctuates due to the perturbation of P_f=0.02. If the number of active molecules remains small, each cluster consists mainly of inactive molecules, and clustering proceeds until the largest cluster is generated. Once the largest cluster is generated, all molecules become active, resulting in a spike in the number of active molecules. Because the enzyme accelerates the inactivation of molecules, the number of active molecules that rapidly increased is also rapidly decreased. Consequently, the large number of active molecules is not sustained for a long period and is generated in the form of spikes.

It is easy to understand why the enzyme does not contribute to the generation of spike trains under no-noise conditions. Under no-noise conditions, the number of active molecules is either 0 or N(i.e., the number of molecules), which never meets the conditions for administering the enzyme.

Figure 5 shows the effect of the enzyme in accelerating the activation of molecules. Due to the enzyme, the number of active molecules is maintained at a higher level and fluctuates perpetually. As long as the number of active molecules remains high, de-clustering proceeds. Once all molecules become monomers, they become inactive, resulting in the generation of a downward spike.

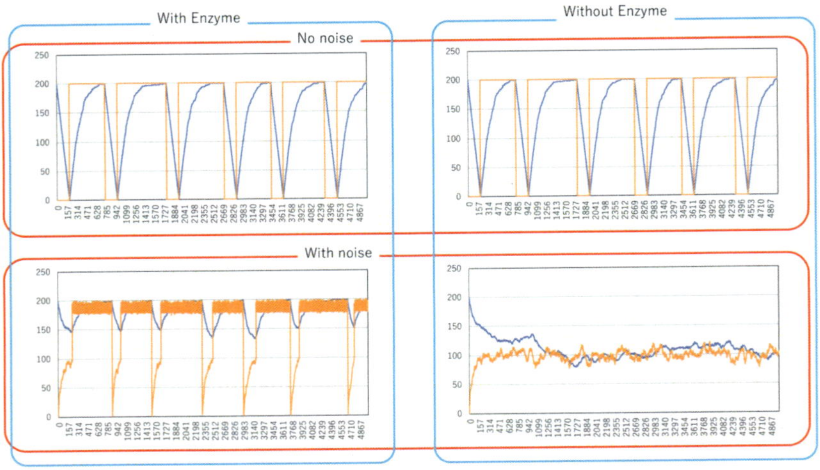

FIGURE 5
The effect of the enzyme to accelerate activation of molecules. Blue lines represent a time series of the number of clusters, and orange lines represents a time series of the number of active molecules.

When the enzyme is administered under perturbation, the population of molecules can generate spike trains. In other words, spike trains are generated at the critical state. This suggests that computation using spike trains could reveal unconventional computing, breaking the trade-off between computational universality and computational efficiency. To verify this, a cellular automaton based on spike trains is implemented in the next section.

3 CHEMICAL CELLULAR AUTOMATA BEYOND TRADE-OFF PRINCIPLE

3.1 Automata and chemical reaction in cells

Cellular automata can be implemented through the interaction between chemical reactions and their environments. This is achieved by introducing a threshold value to the number of activated molecules in a chemical reaction, discretizing it, and using it as a value for a cellular automaton. Since a chemical reaction changes its value depending on small perturbations, an automaton implemented using a chemical reaction can be considered an automaton with inherent fluctuations. This type of automaton has different properties from other automata, such as deterministic ones that follow a universal deterministic rule, and stochastic ones that follow a given probability distribution. Here, we refer to cellular automata implemented through chemical reactions as chemical-implemented automata.

Chemical-implemented automaton are implemented to mimic a given deterministic automaton. First, we define a deterministic cellular automaton. The cellular automata in this section are defiend by $< N, \mathbf{B}, f : \mathbf{B} \times \mathbf{B} \rightarrow \mathbf{B} >$ where N represents the number of cells (i.e., system size), $\mathbf{B}$ represents the set for cell state and $\mathbf{B}=\{0, 1\}$, and f represents the transition rule for the state of a cell. Transition rule of the cellular automaton is here expressed as

$$a_i^{t+1} = f(a_{i-1}^t, a_i^t). \tag{20}$$

Just like ECA (elementary cellular automata), we define the rule number as,

$$r = \sum_{k=0}^{4} d_k 2^k, \tag{21}$$

where $d_k = f(a, b)$ and $k = 2a + b$ with $a, b \in \{0, 1\}$.

Based on a given cellular automaton, the chemical-implemented cellular automaton is defined by the following. Given a cellular automaton of $< N, \mathbf{B}, f : \mathbf{B} \times \mathbf{B} \rightarrow \mathbf{B} >$, Kchambers are given for chemical reaction. In the each chamber, N molecules are initially prepared under a specific perturbation of p_f mentioned before. The time step of the chemical-implemented cellular automaton is represented by $t(\tau)$. The partition of molecules of the ith chamber at the τth step is expressed as

$$P_i^{t(\tau)} = \{\{m_i^{t(\tau)}(a), ..., m_i^{t(\tau)}(b)\}, \{m_i^{t(\tau)}(c), ..., m_i^{t(\tau)}(d)\}, ..., \{m_i^{t(\tau)}(e), ..., m_i^{t(\tau)}(f)\}\}, \tag{22}$$

where $a, b, ..., f \in \{0, 1, ..., \}$ and $a \neq b \neq ... \neq f$. Elements of a partition represent clusters and are designated as follows.

$$C_i^{t(\tau)}(1) = \{m_i^{t(\tau)}(a), ..., m_i^{t(\tau)}(b)\}, \tag{23}$$

$$C_i^{t(\tau)}(2) = \{m_i^{t(\tau)}(c), ..., m_i^{t(\tau)}(d)\}, ..., \tag{24}$$

$$C_i^{t(\tau)}(s_i^{t(\tau)}) = \{m_i^{t(\tau)}(e), ..., m_i^{t(\tau)}(f)\}. \tag{25}$$

The value of $s_i^{t(\tau)}$ represents the number of clusters of the ith chamber at the τth step, such that

$$s_i^{t(\tau)} = \#P_i^{t(\tau)}. \tag{26}$$

The number of active molecules in the ith chamber at the $t(\tau)$th step is expressed as

$$a_i^{t(\tau)} = \#\{m_i^{t(\tau)}(k) \in M | v(m_i^{t(\tau)}(k)) = 1\}. \tag{27}$$

where $M = \{m_i^{t(\tau)}(a), ..., m_i^{t(\tau)}(f)\}$.

In each chamber chemical reaction proceeds by receiving the noise, clustering, de-clustering, activation and inactivation. At first, the state of molecules is flipped with probability, p_f, where it is set by p_f=0.4 for the chemical automata. Receiving the noise is accomplished under p_f=0.4 as

$$v(m_i^{t(\tau)}(k)) = (v(m_i^{t(\tau)}(k)) + 1) mod 2. \tag{28}$$

The clustering is accomplished in each chamber. Two clusters (i.e., elements of partition), $C_i^{t(\tau)}(k)$ with $k = w$ or $k = q$ and $w \leq q$ and $w \neq q$ are randomly chosen and if the clusters satisfy the condition such that

$$\#\{m_i^{t(\tau)}(k) \in C_i^{t(\tau)}(k) | v(m_i^{t(\tau)}(k)) = 1\} / \#C_i^{t(\tau)}(k) \leq p_{cl} \tag{29}$$

and $s_i^{t(\tau)} \neq 1$ then the clusters are united into one cluster by

$$C_i^{t(\tau)}(w) = C_i^{t(\tau)}(w) \cup C_i^{t(\tau)}(q). \tag{30}$$

such that for left-hand term,

$$C_i^{t(\tau)}(w) = \{m_i^{t(\tau)}(w_1), ..., m_i^{t(\tau)}(w_m), m_i^{t(\tau)}(q_1), ..., m_i^{t(\tau)}(q_n)\} \tag{31}$$

where for right-hand term,

$$C_i^{t(\tau)}(w) = \{m_i^{t(\tau)}(w_1), ..., m_i^{t(\tau)}(w_m)\}, C_i^{t(\tau)}(q) = \{m_i^{t(\tau)}(q_1), ..., m_i^{t(\tau)}(q_n)\}. \tag{32}$$

As the allocation of $C_i^{t(\tau)}(q)$ becomes vacant, the allocation for q-cluster will be empty, and the indexes of the clusters after q will be shifted forward such that

$$C_i^{t(\tau)}(x) = C_i^{t(\tau)}(x + 1) \tag{33}$$

for $x \geq q$.

The de-clustering is defined as the following. One cluster (i.e., an element of partition), $C_i^{t(\tau)}(h)$ is randomly chosen and if the cluster satisfies the condition such that

$$\#\{m_i^{t(\tau)}(k) \in C_i^{t(\tau)}(k) | v(m_i^{t(\tau)}(k)) = 1\} / \#C_i^{t(\tau)}(k) \geq 1 - p_{cl} \quad (34)$$

and $s_i^{t(\tau)} \neq \#M$ (i.e., all molecules are not monomers) then the cluster is divide into two clusters such that after de-clustering,

$$\begin{aligned} C_i^{t(\tau)}(h) &= \{m_i^{t(\tau)}(h_1), ..., m_i^{t(\tau)}(h_m)\}, \\ C_i^{t(\tau)}(s_i^{t(\tau)}) &= \{m_i^{t(\tau)}(h_m + 1), ..., m_i^{t(\tau)}(h_n)\} \end{aligned} \quad (35)$$

where before de-clustering

$$C_i^{t(\tau)}(h) = \{m_i^{t(\tau)}(h_1), ..., m_i^{t(\tau)}(h_m), m_i^{t(\tau)}(h_m + 1), ..., m_i^{t(\tau)}(h_n)\}, \quad (36)$$

and the value of h_m is randomly determined. It implies that newly constructed cluster is allocated for the end of the line.

Activation and inactivation of the chemically-inspired automaton is simply defined by the following. If $s_i^{t(\tau)} = 1$ for any $k \in \{1, 2, ..., \#M\}$

$$v(m_i^{t(\tau)}(k)) = 1. \quad (37)$$

If $s_i^{t(\tau)} = \#M$ (i.e., all molecules are monomers) for any $k \in \{1, 2, ..., \#M\}$

$$v(m_i^{t(\tau)}(k)) = 0. \quad (38)$$

After clustering, de-clustering, activation and inactivation, all states are updated for any $k \in \{1, 2, ..., \#M\}$

$$m_i^{t(\tau+1)}(k) = m_i^{t(\tau)}(k), \quad (39)$$

and for any $j \in \{1, 2, ..., s_i^{t(\tau)}\}$

$$P_i^{t(\tau+1)} = P_i^{t(\tau)}. \quad (40)$$

With these updates, both the unchanged and changed states are updated. Depending on these updates, the number of activated molecules and the number of clusters are also updated.

In each chamber, clustering, de-clustering, activation, and inactivation proceed independently of each other. Only through the addition or removal of the enzyme can a chamber interact with the nearest chamber, thereby revealing a chemical cellular automaton.

3.2 Chemical-inspired Automata and their performance

Corresponding to an initial condition for a cellular automaton which is expressed as a binary array of $\mathbf{B}^N$, the enzyme is either administrated or not in each chamber.

If $a_i^t = 1$, then the enzyme to accelerate inactivation is administrated, and otherwise, the enzyme is not given. They are defined by the following.

$$(a_i^t = 1) \Rightarrow ((0.7 \geq r_0 \geq 0.5) \Rightarrow v(m_i^{t(1000 \leq \tau \leq 2000)}(k)) = 0) \tag{41}$$

where $r_0 = \#\{m_i^{t(\tau)}(k) \in C_i^{t(\tau)}(k) | v(m_i^{t(\tau)}(k)) = 0\}/N$. It implies that if the number of inactive molecules is larger than that of active molecules then the inactive molecules are increased, and that shows the acceleration of inactivation of molecules. If $a_i^t = 0$, the enzyme is not administrated.

The spike is identified and counted by

$$Count_i^t = Count_i^t + 1 \tag{42}$$

if the spike is identified such that

$$(s_i^{t(\tau)} \geq 0.95N) \wedge (s_i^{t(\tau)} \geq s_i^{t(\tau-1)}) \wedge (s_i^{t(\tau)} \geq s_i^{t(\tau+1)}) \tag{43}$$

as far as $t(1000 \leq \tau \leq 2000)$. This implies that $s_i^{t(\tau)}$ is the maximal point against τ, and that the number of active molecules generates a spike at $t(\tau)$. After the chemical reaction in $t(1000 \leq \tau \leq 2000)$, and i.e., at $\tau = 2000$, the number of the spikes is determined as either 0 or 1 such that

$$(Count_i^t \leq \theta_{cout}) \Rightarrow (a_i^t = 1) \tag{44}$$

otherwise, $a_i^t = 0$. In other words, the state of cellular automaton, a_i^t is modified via chemical reaction. After that, by using a cellular automaton rule such as

$$a_i^{t+1} = f_r(a_{i-1}^t, a_i^t). \tag{45}$$

The next state of chemical reaction in the ith chamber is updated such that for any $k \in \{1, 2, ..., \#M\}$

$$m_i^{(t+1)(\tau=0)}(k) = m_i^{t(\tau=2000)}(k). \tag{46}$$

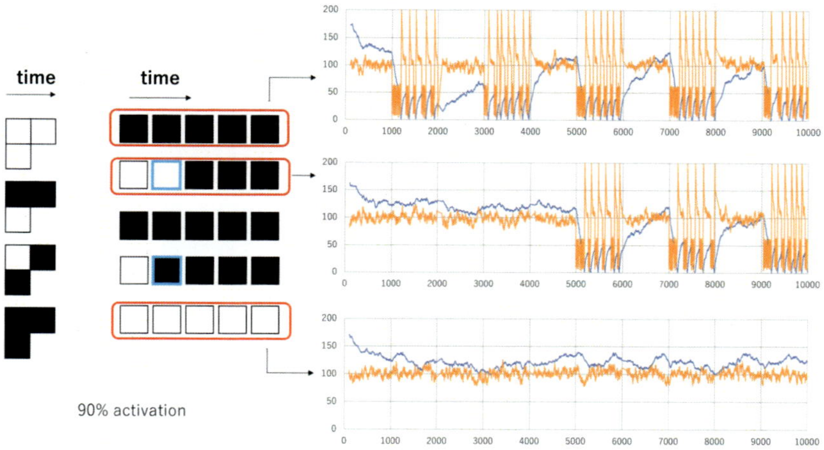

FIGURE 6
Example of chemical-inspired automaton. Left four patterns of the transition represents a rule of cellular automaton; left two squares represent a_{i-1}^t, a_i^t, right one square represents a_i^{t+1}. Central five arrays of squares represent the state of a_i^t, where a black square represents the state of 1, and blank represents the state of 0. A square surrounded by blue loop represents the state which does not follow the transition rule of cellular automaton. Right three graphs represent the number of clusters (blue) and of active molecules (orange) plotted against time step expressed by $t * \tau$

and for any $j \in \{1, 2, ..., s_i^{t(\tau)}\}$

$$P_i^{(t+1)(\tau=0)} = P_i^{t(\tau=2000)}. \tag{47}$$

In the chemical-inspired automaton, the interaction between the chambers corresponding to cells can be implemented by the automaton transition rule. The state of each cell is modified by chemical reactions involving clustering, de-clustering, activation, inactivation, and acceleration of inactivation in each chamber. As a result, chemical reactions contribute to unconventional computing, differing significantly from deterministic cellular automata.

Figure 6 shows the time development of a cellular automaton with $r = 14$, where five cells are coupled with boundary conditions set to all 0. It is clear that if the number of spikes exceeds the threshold value, the discrete state is determined as 1; otherwise, it is 0. An array consisting of black and white squares surrounded by a red loop corresponds to the time series of the number of active molecules and clusters in the chemical reaction. Note that the state of the cellular automaton is sometimes not consistent with the rule; in such cases, the square representing the state of a cell is surrounded by a blue square in Figure 6.

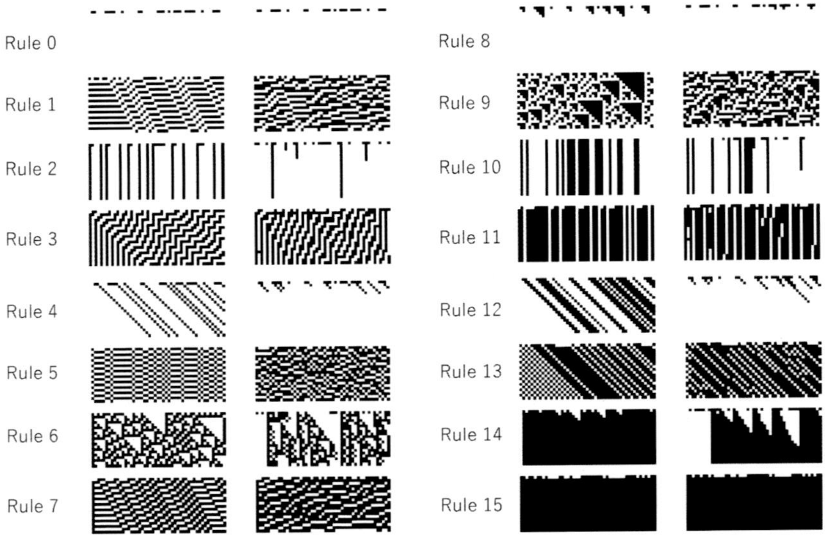

FIGURE 7
A pair of pattern development of conventional cellular automaton (left) and of chemical-inspired cellular automaton. Black squares represent the state, 1, and blanks represent the state, 0. Accompanied rule number represents the transition rule of both conventional and chemical-inspired cellular automaton.

Figure 7 shows a pair of time developments generated by usual cellular automata (left) and by chemical-inspired cellular automata (right), with the transition rule provided by the accompanying rule number. The horizontal and vertical lines represent an array of cells (chambers) and time steps, respectively, where black squares and blanks represent the 1 and 0 states, respectively. Since some cells do not follow a given transition rule, the patterns generated by chemical-inspired cellular automata differ from those produced by usual cellular automata. These differences could contribute to the advantages of chemical-inspired cellular automata.

There are many attempts to implement unconventional computing using chemical and/or biological devices (e.g. [39–42]. As these devices are dynamic and unstable, they may show less accuracy in computation. However, biological devices, in particular, could demonstrate robust computing because they have redundant computing mechanisms and/or the ability to self-repair. These abilities are inherently connected to instability. This raises the question of how unstable interactions play a role in computing.

Michael Conrad argued that unstable interactions could contribute to computation, especially regarding the trade-off between computational universality and computational efficiency [38]. He originally considered a population of biochemical molecules as a computer, interpreting the combinations of molecules through self-aggregation as computation. If a population of molecules has an affinity to all other molecules (population A), it can reveal all possible combinations. However, since any combination can be achieved by chance and/or molecular diffusion, it can take a long time to reach the desired combination. In contrast, if a population of molecules has an affinity only to a specific type of molecule (population B), only very specific combinations are generated, albeit very quickly. The number of realized molecular combinations can be compared to computational universality, and the average time to reach a reachable combination can be compared to computational efficiency. Population A shows high computational universality and low computational efficiency, while population B shows low computational universality and high computational efficiency, suggesting a trade-off principle between computational universality and efficiency.

Although Conrad's trade-off principle suggests the advantage of general unconventional computing, it appears speculative. In the field of cellular automata, one of the authors verified a strict trade-off principle between computational universality and efficiency in elementary cellular automata [35, 36]. Given all possible initial configurations of a binary sequence, computational universality is defined by the number of reachable configurations divided by the number of all possible configurations, and computational efficiency is defined by the inverse of the average time to reach all reachable configurations. If the inverse of computational efficiency (i.e., the average time to reach all reachable configurations) is plotted against computational universality for each cellular automaton, it is evident that no inverse of computational efficiency can be shortened beyond the minimal time dependent on computational universality. The shorter the minimal time against computational universality (i.e., the higher the computational efficiency), the lower the computational universality, implying a trade-off principle in elementary cellular automata.

While elementary cellular automata are implemented as synchronous automata, it has also been shown that unconventional elementary cellular automata based on asynchronous updating break the trade-off principle found in elementary cellular automata. This can be one explicit advantage of unconventional computing containing implicit instability. Therefore, we estimate whether the cellular automata defined here also show the trade-off between computational universality and efficiency, and evaluate how chemical-inspired cellular automata might break this trade-off principle if it exists.

First, computational universality is defined as follows. Given an n-bit binary sequence, the number of all possible configurations is 2^n. If a configuration is generated by applying a transition rule of the cellular automaton to one of the possible initial configurations within 2^n time steps, the configuration is determined as a reachable configuration, which is an element of a set,

$$S_{reach}(f_r) = \{(b_1, b_2, {}^\wedge f, b_n) \in \mathbf{B}^n | (b_1, b_2, {}^\wedge f, b_n) = f_r^t(a_1, a_2, {}^\wedge f, a_n), \\ \exists(a_1, a_2, {}^\wedge f, a_n) \in \mathbf{B}^n, \exists t \in \{1, 2, {}^\wedge f, 2^n\}\}. \tag{48}$$

If one checks time development of a cellular automaton Rule r for all possible initial configurations in 2^n time steps, one can count the number of reachable configuration represented by $\#S_{reach}(f_r)$, and then the computational universality, $C_{univ}(f_r)$ is expressed as

$$C_{univ}(f_r) = \#S_{reach}(f_r)/2^n. \tag{49}$$

where all possible boundary conditions are given. For a cellular automaton, Rule r, the time to an element of a set of reachable configurations is expressed as

$$t_{reach}(b_1, b_2, {}^\wedge f, b_n; f_r) = \sum_{(a_1, a_2, {}^\wedge f, a_n) \in \mathbf{B}^n} t_{ind}(a_1, a_2, {}^\wedge f, a_n)/2^n \tag{50}$$

where for $(a_1, a_2, {}^\wedge f, a_n)$ in $\mathbf{B}^n$,

$$(b_1, b_2, {}^\wedge f, b_n) = f_r^t(a_1, a_2, {}^\wedge f, a_n) \Rightarrow t_{ind}(a_1, a_2, {}^\wedge f, a_n) = t; \tag{51}$$

$$\forall t \in \{1, 2, .., 2^n\}, (b_1, b_2, {}^\wedge f, b_n) \neq f_r^t(a_1, a_2, {}^\wedge f, a_n) \\ \Rightarrow t_{ind}(a_1, a_2, {}^\wedge f, a_n) = 2^n. \tag{52}$$

The computational efficiency, $C_{eff}(f_r)$, is defined by

$$C_{eff}(f_r) = \Big(\sum_{(b_1, b_2, {}^\wedge f, b_n) \in S_{reach}(f_r)} t_{reach}(b_1, b_2, {}^\wedge f, b_n; f_r)\Big)^{-1}. \tag{53}$$

Figure 8 is an example of cellular automaton showing high computational efficiency and low computational universality. That is rule 0 which generates only a configuration consisting of all 0 for any initial configuration. Therefore, $S_{reach}(f_0)$=(0, 0, 0, 0) and $\#S_{reach}(f_0)$=1, and then $C_{univ}(f_0)=1/2^4=1/16$ that shows low computational universality. Since reachable configuration, (0,0,0,0) can be reached just in 1 time step for any initial configuration, $C_{eff}(f_0)$=1/1=1 that shows high computational efficiency.

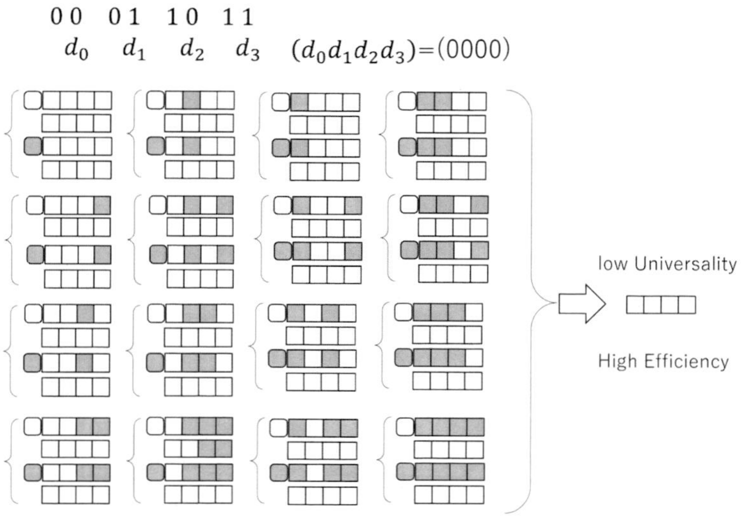

FIGURE 8
Highly efficient and low universal cellular automaton. Each arrays of squares represents time development with boundary condition, gray square represents state 1, and blank represents state 0

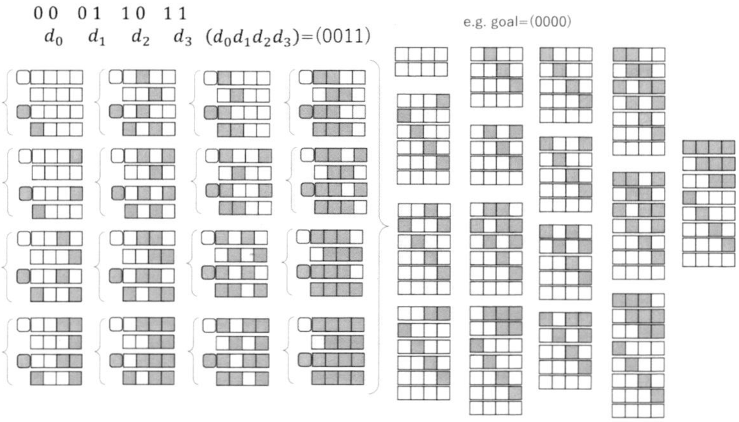

FIGURE 9
High universal and low efficient cellular automaton. Each arrays of squares represents time development with boundary condition, gray square represents state 1, and blank represents state 0. Left diagram shows that any possible 2^n configurations can be obtained by this cellular automaton. Rught diagram shows that the time development toward (0, 0, 0, 0) from any initial configurations.

In contrast, Figure 9 is an example of a cellular automaton showing high computational universality and low computational efficiency. This cellular automaton is indicated by Rule 6. It shows that any 4-

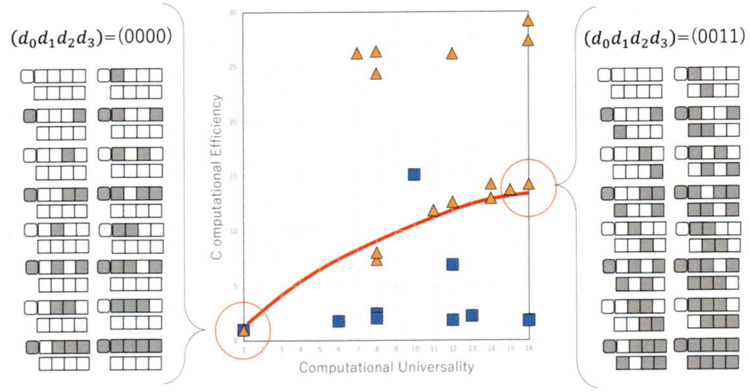

FIGURE 10
Inverse of the computational efficiency against the computational universality of the conventional cellular automata (orange triangle) and of the chemical-inspired cellular automata. Red curve represents the trade-off between the computational universality and efficiency found in conventional cecllular automata.

bit configurations can be generated by applying the rule to possible initial configurations. This implies the greatest computational universality since $\#S_{reach}(f_6)$=2^4=16, resulting in $C_{univ}(f_6)$=$2^4/2^4$=1. The right diagram in Figure 9 shows time developments to a specific configuration (0, 0, 0, 0) from all possible initial configurations. From this set, one can obtain $t_{reach}(0, 0, 0, 0; f_6) = (1 + 6 + 6 + 6 + 4 + 4 + 6 + 7 + 5 + 5 + 5 + 5 + 7 + 7 + 8 + 8)/16 = 90/16$. As this could be similar to the average of all possible $t_{reach}(b_1, b_2, \hat{}f, b_n; f_6)$, $C_{eff}(f_6)$ is nearly equal to $16/90 \leq 1$, which is very low.

4 DISCUSSION AND CONCLUSION

We previously proposed an artificial chemical system where chemical reactions involve repeating self-aggregation and self-dispersion, and we demonstrated the implementation of logic gates using this system. In that work, we suggested the potential for breaking the trade-off between computational universality and computational efficiency. To investigate whether this chemical system truly breaks this trade-off, we implement chemical-inspired cellular automata and compare them with conventional cellular automata.

First, we describe the artificial chemical system with greater mathematical precision than in our previous work. The system consists of chemical molecules whose states alternate between active and inactive. In the absence of perturbation, if all molecules become monomers, they turn inactive, while if all molecules form a single large cluster, they become active. Inactive molecules promote self-aggregation, while active molecules promote self-dispersion, leading the chemical population to repeatedly cluster and de-cluster.

In contrast, under perturbation conditions, the interactions between perturbation, activation, inactivation, and clustering-declustering result in a dynamically fluctuating state characterized by $1/f$ noise. Since $1/f$ noise is associated with phenomena such as the edge of chaos and self-organized criticality, high computational performance is expected. However, instability in chemical reactions or unconventional computing often leads to disadvantages such as errors.

We evaluate computational capability in terms of the trade-off between universality and efficiency. For example, a knife is versatile, capable of tasks such as cutting paper, shaving wood, prying corks, and turning screws, but each task is performed less efficiently compared to specialized tools like scissors, planes, corkscrews, and screwdrivers. This demonstrates a trade-off between high universality and low efficiency.

In living systems, specialists and generalists also illustrate this trade-off. Specialists excel in specific niches but are vulnerable to environmental changes, while generalists, though less specialized, can adapt to various niches and are more resilient. The development of the brain and intelligence in humans exemplifies the ability to break this trade-off, allowing us to excel in diverse tasks.

Under perturbation, administering an enzyme for inactivation can generate controllable spike trains. We implement chemical-inspired cellular automata by configuring multiple chambers where populations of chemical molecules interact through enzyme administration. Our results show that chemical-inspired cellular automata can break the trade-off between computational universality and efficiency found in conventional cellular automata. We believe that this ability to break the trade-off is akin to natural intelligence rather than merely logical intelligence, representing a distinct advantage of unconventional computing.

This research was funded by JSPS 22K12231, 23K00237, and co-financed by Waseda University.

REFERENCES

[1] J. Klijin and B. Engberts. (2005). Fast reaction on water. *Nature*, 435:746–747.

[2] Y. Matsunaga and Y. Sugita. (2020). Use of single-molecule time-series data for refining conformational dynamics in molecular simulations. *Current Opinions in Structural Biology*, 61:153–159.

[3] L. Bacic, A. Sabantsev, and S. Dendl. (2020). Recent advances in single-molecule fluorescence microscopy render structural biology dynamic. *Current Opinions in Structural Biology*, 65:61–68.

[4] N. Kawamoto, H. Ito, I. Tokuda, and H. Iwasaki. (2020). Damped circadian oscillation in the absence of kaia in synechococcus. *Nature communications*, 11:2242.

[5] Y. Furuike, A. Mukaiyama, D. Ouyang, K. Ito-Miwa, D. Simon, E. Yamashita, T. Kondo, and S. Akiyama. (2022). Elucidation of master allostery essential for circadian clock oscillation in cyanobacteria. *Sci. Adv*, 8:eabm8990.

[6] M. Fang, A. Chavan, A. LiWanga, and S. Goldena. (2023). Synchronization of the circadian clock to the environment tracked in real time. *Proc. Natl. Acad. Sci. U.S.A*, 120:e2221453120.

[7] A. Chen and M. Elowitz. (2021). Programmable protein circuit design. *Cell*, 184:2284–2301.

[8] D. Zhang, Y. Cao, Q. Ouyang, and Y. Tu. (2020). The energy cost and optimal design for synchronization of coupled molecular oscillators. *Proc. Natl. Acad. Sci. U.S.A*, 16:95–100.

[9] D. Gillespie. (1977). Exact stochastic simulation of coupled chemical reactions. *J. Phys. Chem.*, 81(25):2340–2361.

[10] D. Gillespie. (2001). Approximated accelerated stochastic simulation of chemically reacting systems. *J. Phys. Chem.*, 115(4):1716–1733.

[11] D. Gillespie. (2007). Stochastic simulation of chemical kinetics. *Annu. Rev. Phys. Chem.*, 55:35–55.

[12] Y. Suzuki, Y. Fujiwara, and J. Takabayashi. (2001). Artificial life applications of a class of p systems: Abstract rewriting systems on multisets. *Lecture Notes in Computer Science*, 2235:35–55.

[13] J.-P. Banarte, P. Fradet, and Y. Radenac. (2006). Generalised multisets for chemical programming. *Math. Struct. in Comp. Science*, 16:557–580.

[14] Y. Gunji and A. Adamatzky. (2024). Computation implemented by the interaction of chemical reaction, clustering, and de-clustering of molecules. *Biomimetics*, 9(7):432.

[15] S. Wolfram. (1983). Statistical mechanics of cellular automata. *Rev. Mod. Phys.*, 55:601–644.

[16] S. Wolfram. (1984). Universality and complexity in cellular automata. *Physica D*, 10:1–35.

[17] S. Wolfram. (2002). *A New Kind of Science*. Wolframscience.com.

[18] P. Bak, C. Tang, and K. Wiesnfeld. (1987). Self-organized criticality: An explanation of 1/f noise. *Phys. Rev. Lett.*, 59:381–384.

[19] P. Bak and C. Tang. (1989). Earthquakes as a self-organized critical phenomenon. *J. Geol. Res.*, 94(15):635–637.

[20] P. Bak and K. Sneppen. (1993). Punctuated equilibrium and criticality in a simple model of evolution. *Phys. Rev. Lett.*, 71:4083–4086.

[21] K. Sneppen, P. Bak, H. Flybjerg, and M. Jensen. (1995). Evolution as a self-organized critical phenomenon. *Proc. Nat. Acad. Sci. USA*, 92:5209–5213.

[22] C. Langton. (1990). Computation at the edge of chaos: Phase transitions and emergent computation. *Physica D*, 42:12–37.

[23] S. Kauffman and S. Johnsen. (1991). Coevolution to the edge of chaos: coupled fitness landscape, poised states, and coevolutionary avalanches. *J. Theor. Biol.*, 149(4):467–505.

[24] S. Kauffman. (2019). *A World beyond Physics. The Emergence & Evolution of Life*. Oxford Univ. Press.

[25] J. Lendner, R. Helfrich, B. Mander, J. Lin, M. Walker, P. Larsson, and R. Knight. (2020). An electrophysiological marker of arousal level in humans. *eLife*, 9:e55092.

[26] M. Santos, J. Szezech, F. Borges, K. Iarosz, K. Caldos, A. Batista, R. Viana, and J. Kurths. (2017). Chimera-like states in a neuronal network model of cat brain. *Chaos, Solitons & Fractals*, 101:86–91.

[27] D. Toker, I. Pappas, J. Lendner, J. Flohrich, D. Mateos, S. Muthukumaraswamy, R. Carhart-Harris, M. Paff, P. Vespa, M. Monti, F. Sommer, R. Knight, and M. DEsposito. (2022). Consciousness is supported by near-critical slow cortical electrodynamics. *Natl. Acad. Sci. U. S. A.*, 119:e2024455119.

[28] C. Kello, G. D.A.B., R. Ferrer-i Cancho, J. Holden, K. Linkenkaer-Hansen, T. Rhodes, and G. Van Orden. (2010). Scaling laws in coginitive sciences. *Trends. Cogn. Sci.*, 14:223–232.

[29] J. Wilting and V. Priesemann. (2019). 25 years of criticality in neuroscience - established results, open controversies, novel concepts. *Curr. Opin. Neurobiol.*, 58:105111.

[30] D. Plentz, T. Ribeiro, S. Miller, P. Kells, A. Vakili, and E. Capek. (2021). Self-organized criticality in the brain. *Front. Phys*, 9:365.

[31] M. Cook. (2004). Universality in elementary cellular automata. *Complex System*, 15:1–40.

[32] Y. Gunji. (1990). Pigment color patterns of molluscs as autonomy, generated by asynchronous automata. *BioSystems*, 23:317–334.

[33] Y. Gunji. (2015). Extended self-organized criticality in asynchronously tuned cellular automata. In B. V. Nicolis, G., editor, *Chaos, Information Processing and Paradoxical Games*, pages 411–429. World Scientific.

[34] D. Uragami and Y. Gunji. (2018). Universal emergence of 1/f noise in asynchronously tuned elementary cellular automata. *Complex Systems*, 27399:414.

[35] Y. Gunji and D. Uragami. (2020). Breaking of the trade-off principle between computational universality and efficiency by asynchronous updating. *Entropy*, 22(9):1049.

[36] Y. Gunji and D. Uragami. (2021). Computational power of asynchronously tuned automata enhancing the unfolded edge of chaos. *Entropy*, 23(11):1376.

[37] D. Uragami and Y. Gunji. (2022). Universal criticality in reservoir computing using asynchronous cellular automata. *Complex Systems*, 31:104–121.

[38] M. Conrad. (1983). *Adaptability, The Significance of Variability from Molecule to Ecosystem*. Plenum Press. New York.

[39] A. Adamatzky. (2010). *Physarum Machines, Comouters from Slime Mould.* World Scientific.

[40] A. Adamatzky, M. Tegelaar, H. Wosten, A. Powell, A.L. Beasley, and R. Mayne. (2020). On boolean gates in fungal colony. *BioSystems*, 193-194:104138.

[41] A. Adamatzky. (2022). Language of funji derived from their electrical soiking activity. *R. Soc. Open Sci.*, 9:211926.

[42] P. Mougkogiannis and A. Adamatzky. (2023). Logical gates in ensembles of protenoid microsphere. *Plos One*, 18(9):e0289433.

Computing with Autonomous Agents on Graphs

DAN V. NICOLAU, JR[1]

[1]*Peter Gorer Department of Immunobiology, School of Immunity and Microbial Sciences, Faculty of Life Sciences and Medicine, King's College London, Guy's Hospital, Great Maze Pond, London, United Kingdom*

Received: September 1, 2024. Accepted: September 28, 2024.

In ongoing work combining theory, simulation and biological experiments, we experimentally demonstrated how self-propelled molecular "agents" moving stochastically through a specially designed nanofabricated network can solve an NP-complete problem (SUBSET SUM) in a massively parallel fashion and, theoretically, in polynomial time. Here, we formally introduce a new computing model, of the distributed parallel type, based on this principle. The computing device, which we call an "N-system", consists of (a) an appropriately designed digraph encoding a specific instance of a decision problem and (b) a population of agents that evolve on the digraph by jumping between connected nodes at each step of the computation, starting at one or more "input nodes". The computation terminates when the occupancy status for all nodes is stable, or after some maximum time has elapsed, at which point the result of the computation is completely described by the occupancy status of one or more "output nodes". The evolution takes place in parallel for all agents that are able to evolve at each step. We show, with an example, how such a system can enumerate the natural numbers as well as how to solve SUBSET SUM in general with this approach. We consider the computational power of this system, describing the power and limitations of N-systems without "stickers" (i.e. without information about agents paths) as well as briefly considering the additional power conferred by stickers. Finally, some issues connected with the physical implementation of N-systems are discussed, as well as their similarities with neural networks, some open problems and future research directions.

Keywords: Networks-based biocomputation, NP-complete problems, unconventional computing

*Corresponding author email: dan.nicolau@kcl.ac.uk

1 INTRODUCTION

Beginning in the late 1960s, it was recognised that there are mathematical problems for which there appear to be no efficient algorithms on conventional computers: that is to say, in the worst case, the computational time grows superpolynomially (if not exponentially) with the instance size of the problem. The formal statement of this conjecture, usually referred to as P vs NP, is one of the most important — and perhaps indeed the most important [1] — in contemporary science. Increasing conviction that so-called NP-complete problems (of which there are now thousands known) indeed will require superpolynomial time on sequential machines and are therefore fundamentally intractable by conventional computers has sparked a deep interest in alternative, massively parallel computing devices and models. In particular, if P != NP, as is widely believed, then only computational models that are "massively parallel", trading superpolynomial time for another superpolynomial resource (e.g. number of DNA strands), can hope to solve NP-complete problems in practice.

Separately, electronic computers, whose extraordinary increase in power over the last decades has followed Moore's law, are now reaching their limits [41]. These limits are a result of various physical constraints, including electronic circuit feature size [54], heating and energy consumption [63]. In this context, there has been increasing interest in developing novel computing technologies that may overcome these barriers (as well as those due to problem "hardness").

Amongst the best-known attempts to develop massively parallel computing models are quantum computing [27], DNA computing [65], P-systems and related technologies [59] and microfluidic computing [16]. Many of these models rely on the natural ability of living things to execute tasks in a totally distributed, asynchronous, parallel manner. DNA computing takes advantage of a huge population of DNA strands to combine at random, effectively generating all possible candidate solutions in one step. P-systems and similar efforts aim to compute by interpreting biochemical reactions and cellular division as computational actions. Unfortunately, these efforts have all been hindered not by theoretical limitations but by "engineering" issues that interfere with the respective technology scaling enough to solve problems of practical size, or, alternatively, to challenge the power of electronic computers[64].

In ongoing work [49] we introduced a new type of parallel computational device that uses a population of autonomous, motile, self-propelled agents (driven by molecular motors, in the implementation we described). The agents explore specially designed, nanofabricated networks whose structure encodes a combinatorial (NP-complete, though not necessarily) mathematical problem of interest. The trajectories of the agents, which are stochastic, represent different candidate solutions to the problem instance, while their positions within the network after some elapsed time represent the final solution to the problem. This technology can be demonstrated to use orders of magnitude less energy per computational step than electronic computers while computing in a massively parallel manner [49]. It is, therefore, an attractive candidate as model of computation that can practically solve NP-complete problems, not least because, by virtue of using existing molecular or cellular components of living systems, much of the engineering "hardware" has been evolutionarily perfected.

While this approach seems promising for these and other reasons, there has not, to date, been a formal statement of its properties in directly mathematical terms and, consequently, no theoretical analysis of its potential computational power and limitations, respectively. This paper fills part of this gap. It considers a simplified model of these network-driven devices, which we here call N-systems, in which the number of agents available for a given population is unlimited, in the specific sense that the number of agents is so large that their density per visited node does not decrease as the computation proceeds. Furthermore, we assume, as in the experimental demonstrations to date, that the agents cannot keep an interrogatable record of their trajectories through the networks, though at the end of the paper we briefly consider such systems as well.

2 NETWORK STRUCTURES

This paper describes an idealised model of a new computing technology of the parallel, distributed type that is based on the idea of converting mathematical problems to networks (that is, directed graphs) and searching through the network for the existence of a path that corresponds to a/the solution to the original problem. We will focus here on combinatorial problems (especially NP-complete or other combinatorial problems), since we are ultimately interested in exploiting the parallel nature of biological systems to solve problems intractable with classical computing devices. Of course, these results apply to problems in other computational classes and any speedup obtained with our parallel devices will carry over to other computational tasks (though not necessarily in a straightforward way).

We will also concentrate, for simplicity, on decision problems, that is, those formulated as YES/NO questions on an infinite set of inputs, e.g. the problem Satisfiability (SAT), which asks, given a propositional formula F in k literals, whether F is satisfiable by some assignment τ of values to the literals or not, i.e. the function computing the decision problem should return YES if such an assignment exists and NO otherwise. Note, however, that for each decision problem in NP (i.e. the problems of most concern to us here), there exists a corresponding search problem. Specifically, given a string x, the search problem consists of finding a string y satisfying a checking relation $R(x, y)$ for the problem if such a string exists; or return NO if x is a NO instance to the problem (y is said to be a "certificate" for x). For NP-complete problems, a search problem can be reduced to the corresponding decision problem in polynomial time on classical sequential devices (if P = NP, then naturally the associated search problem also has a polynomial-time algorithm). For example, an algorithm for SAT can be used to find a truth assignment τ satisfying the propositional formula F by setting each literal in F to TRUE or FALSE depending on which assignment keeps F satisfiable (which we can determine with our algorithm). Thus, although we will, for reasons of practical interest, concentrate on the decision versions of combinatorial (and specifically, NP-complete problems), the design, computing power and usefulness of our devices, as will be seen, are unaffected by these choices in general.

We start by introducing the "hardware" component of our computational devices: digraphs whose design reflects a particular problem of interest. First, consider an arbitrary decision problem D. Then we have the following definitions.

Definition. Let D_X be an instance of D. Then $|D_X|$ denotes the size of D_X, that is, the number of bits required to represent D_X. For example, for D_X being an instance of the Subset Sum Problem, $|D_X|$ is the number of bits needed to represent the set of integers and the target integer, respectively.

Definition. Let E be a weighed digraph. Then $|E|$ denotes the size of the digraph E, that is, the minimum number of bits needed to represent the adjacency and edge weights matrix.

We are now ready to define formally the relationship between decision problems and digraphs that the rest of the paper relies on.

Definition. A digraph E_X ***encodes*** a given instance D_X of D if and only if, given two disjoint sets of input nodes $I = \{I_1, I_2, \dots I_k\}$ and output nodes $\Omega = \{\Omega_1, \Omega_2, \dots \Omega_l\}$ in E_X, there exists at least one loop-free path p between any node in I and any node in Ω. We will denote this relationship by the notation $D_X \rightarrow E_X$.

Intuitively, we are looking for a class of digraphs E that somehow captures the structure of the decision problem in such a way that, for every instance D_X of D, we can find a digraph E_X in which there is a way to get from one or more (predefined) input nodes to one or more (predefined) output nodes if and only if the decision problem on D_X (as defined above) can be solved in the affirmative. Conversely, we need to guarantee that such a path cannot be found if the decision problem on D_X has a negative answer.

Before illustrating this concept of digraph encodings with a few simple examples, we will make two more definitions (that essentially recapitulate their counterparts in Boolean and arithmetic circuits [ref]) and a few observations.

Definition. For a digraph E_X, the **fanin** is the maximum number of edges leading to any node in E_X from its neighbours. The ***fanin*** of a node j is the number of edges leading to j from its neighbours.

Definition. For a digraph E_X, the ***fanout*** is the maximum number of edges leading away from any node in E_X to its neighbours. The ***fanout*** of a node j is the number of edges leading from j from to each of its neighbours.

Observe that we can trivially guarantee the existence of an encoding for each problem instance, since for each such instance D_X, we can simply solve D_X by other means and set $D_X \rightarrow E_0$ where E_0 is a 2-node digraph with one input node ι_0 and one output node ω_0 and only one edge, pointing from ι_0 to ω_0 if D_X has at least one solution and omit the edge if it has no solutions. Obviously, this would in no way help to solve the combinatorial problem of interest efficiently. For practical purposes, what we would like to design are non-trivial networks that encode a given problem instance without solving the instance directly. Specifically, what we would like to have is an algorithm A, such that, given an instance D_X of D, A generates a conversion $D_X \rightarrow E_X$ (note that these need not be unique) in polynomial (i.e. $|D_X|$ to some fixed power) time.

Secondly, note that without loss of generality, we can limit our discussion to encodings with exactly one input and one output node, since for a given encoding E_X with multiple input and output nodes, $I = \{I_1, I_2, \ldots I_k\}$ and $\Omega = \{\Omega_1, \Omega_2, \ldots \Omega_l\}$, we can simply add (a) a new input node I_0 with fanin 0 and an edge from I_0 to each I_i for $i = 1, 2, \ldots, k$ and (b) a new output node Ω_0 with fanout 0 and an edge from Ω_0 to each Ω_i for $i = 1, 2, \ldots, l$. Intuitively, this just expresses the idea that we can "tie" together multiple input nodes to one source, and "funnel" multiple output nodes into one exit from the network. For the rest of the paper, we will focus on this special case of encodings, noting that multiple inputs and outputs respectively are, however, expected to be of practical value in physically implementing the computing devices described in the next section by potentially allowing higher agent throughput and post-processing, respectively [ref Interface].

A simple – but not trivial – example of an encoding of this type can be imagined immediately for the problem of establishing node-to-node connectivity in a digraph. Given a digraph G and two nodes s and t, the s-t connectivity problem (STCONN) is defined by

$$\text{PATH} = \{\langle G, s, t\rangle \mid G \text{ is a directed graph with a path from vertex } s \text{ to } t\} \quad (1)$$

It is immediately obvious that $G \rightarrow G$, with $I = s$ and $\Omega = t$ and therefore, the decision problem STCONN encodes itself for each instance. We will say that encoding is *invariant* under the choices of problem STCONN (at a very high level, similarly to e.g. the exponential function e^x being invariant under differentiation).

We are naturally interested in how the "size" of encodings grows with problem instance size. Specifically, because our computing machines will combine physical digraph encodings with populations of exploratory agents, we are specifically concerned with the number of unique paths to be explored. Considering an instance D_X of D, we can define the number of unique paths from I to Ω by $u(E_X)$ for a digraph E_X that satisfies $D_X \rightarrow E_X$. Then given a sequence of instances D_i and encodings E_i with $D_1 \rightarrow (E_1, F)$, $D_2 \rightarrow (E_2, F)$, ... we can denote by $U_E(|D_i|)$ the *worst (highest) case growth* of U_E with the input size $|D_i|$ for that problem instance; that is,

$$U_E(|D_i|) = \max\{u(E_i) \mid D_i \rightarrow E_i\} \quad (2)$$

The usefulness of U_E lies mainly in potentially allowing us to classify encodings. Since the autonomous agents explore physical realisations of the encodings stochastically, the number of agents must certainly be larger than u_E in order for all of the unique paths to be visited. For decision problems where we cannot find encodings whose U_E grows slowly (polynomially) with $|D_i|$, intuitively, we need to employ a population of agents whose size grows superpolynomially with $|D_i|$ (below, we restate this more formally in relation to the classical complexity classes) in order to sample, i.e. explore, the whole solution space in a relaible way.

3 TRANSITION N-SYSTEMS

We are now ready to introduce the main subject of our investigation: a computing mechanism, which we here call an "N-system", designed as a distributed parallel machine, having as the underlying structure digraph encodings of decision problem instances, as defined above. The primary additional feature, which is indispensable for the the computing process, is a population of self-propelling "agents" that "evolve", in the sense of exploring paths through the encoding digraph in a parallel, stochastic manner. The computation terminates if either an output node is reached (solving the problem instance in the affirmative) or if no more transitions are possible (solving it in the negative).

In this paper, we consider a special case of these systems with the property that, at each step of the computation, all nodes that contain agents transmit agents to all of their neighbours with a negligible loss of agents at the source node. This corresponds essentially to a situation where the system has an unlimited number of agents initially placed at the source node, so that their density does not decrease as they percolate through the digraph encoding. Although realistic systems, such as the one described in Nicolau et al. (2016) clearly have finite (though large) numbers of agents available, this simplification allows us, for the time being, to focus on the computing power of N-systems as a function of encoding efficiency.

Definition. An N-system M of order m is formally defined by $M = (E, I, \Omega)$, where E is a digraph of m nodes described by a m x m adjacency matrix W and I and Ω are the input and output nodes as defined above. Every node (junction) can be in one of two possible states, either 1 or 0, corresponding to being populated or not by agents, respectively. The state of node i at time t is denoted by $V_i(t)$, with $V_I(0) = 1$ and $V_\Omega(0) = 0$. The state of the N-system M at time t is the vector $V(t)$. The next state of a node is computed by

$$V'' i(t+1) = sgn(H_i(t)) \tag{3}$$

where

$$H'' i(t) = \sum_{j=1}^{m} W_j i(t) V_j(t) \tag{4}$$

The next state of the N-system M, i.e., $V(t+1)$, is computed from the current state by performing the evaluation (4) at at all the nodes of E simultaneously. We say that M is *stable* at time t iff its state $V(t)$ satisfies $V(t+1) = sgn[WV(t)]$, i.e. if no more changes in the state of M are possible and all nodes that can be visited by agents have been visited. The computation terminates when M is stable. Note, however, that for the purposes of deciding an underlying problem instance $D_X \rightarrow E_X$, if $M = \{E_X, I, \Omega\}$, an affirmative answer corresponds to the smallest t such that $V_\Omega(t) = 1$.

In principle, one could define three modes of operation: (a) the sequential mode, in which the evaluation is carried out at only one node in any time interval, (b) a quasi-parallel mode, in which it is carried out at a partial set of nodes in each interval and (c) a parallel mode, in which it is evaluated at all nodes in each interval. Clearly, classical (sequential) computer simulations of an Nsystem are going to take place in either the serial mode or a quasi-parallel mode (if multiple processors are available). However, since we are interested in building N-systems using biological, self-propelled agents, and since these operate concurrently and independently (as in Nicolau et al., 2016), we focus our attention on the fully parallel case.

The general flow of computing processes on N-systems is illustrated in Figure 1. The input node (or nodes) in an encoding of a problem instance are connected to the output node (or nodes) by a set of *internodes* and associated edges. A large population of physical agents starts at the input node (or nodes) I_0. At $t = 1$, each agent transitions to one of its neighbours, chosen at random. Their positions and trajectories are updated according to equations 3-4 above. If any agents have reached the output node(s), we conclude the decision problem can be answered in the affirmative and may terminate the computation

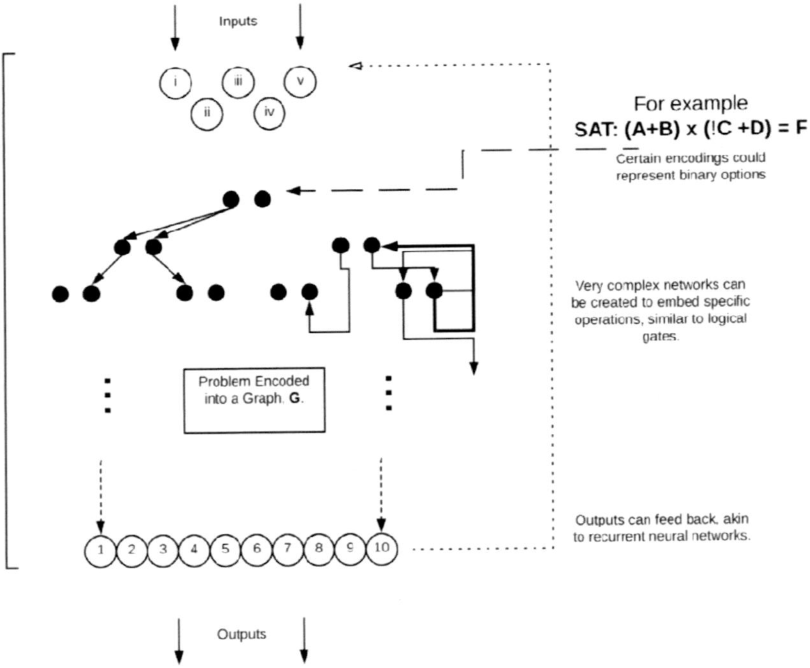

FIGURE 1
Schematic of general computational processes on N-systems. Agents start at the input node(s) and percolate stochastically through the encoding of a problem instance by the simultaneous application of transition rules between nodes, emerging at the output node(s) where the trajectory function F is evaluated. If F is satisfied for any agent or if no more transition rules can be applied (e.g. because all agents have reached nodes with fanout 0, be they output or other nodes), then the computation stops.

(or stop looking). To ensure no solution exists, we simply wait until the system has reached a stable state as described above.

Before returning to the issue of the kind of computing power that N-systems can deploy, we first wish to illustrate how N-systems might in practice solve specific combinatorial problems via a simple example.

4 EXAMPLE: SUBSET SUM PROBLEM

To demonstrate how an N-system might compute a hard function in practice, we can turn to a geometrically intuitive encoding for SUBSET SUM, a benchmark classical NP-complete problem [25]. This problem asks whether, given a set $S = \{s_1, s_2, ..., s_N\}$ of N integers, there exists a subset of S whose elements sum to a target sum, T. To find all possible subset sums by exploring all possible subsets requires the testing of 2^N different combinations, which—even for modest values of N—is clearly impractical on electronic computers because of exponentially increasing time requirements. Although more sophisticated algorithms exist [13, 35, 57], none of these avoids the exponentially growing exploration time, a property that is harnessed in some cryptography systems to generate encoded messages [33].

SUBSET SUM can be formally encoded in the sense described in the previous section by interpreting (a) transitions of agents between nodes as elementary operations of addition and subtraction in unary and (b) their spatial positions in the network as"running sums". The basic principle is illustrated in Fig 2 below for a small subset $S = \{1, 3, 5, 11\}$. The digraph encoding takes a triangular form (in a planar encoding) in which downward transitions of agents correspond to skipping a number in the S while diagonal transitions correspond to adding that integer to a running sum. The input node is always the top left, and all other nodes are assigned $V_i(0) = 0$. The "vertical distance" (measured as the number of junctions) between two subsequent rows of nodes represents an integer from S. The transition of agents downward from a given split junction is equivalent to specifically excluding the corresponding integer from the summation, whereas diagonal transitions are equivalent to its inclusion.

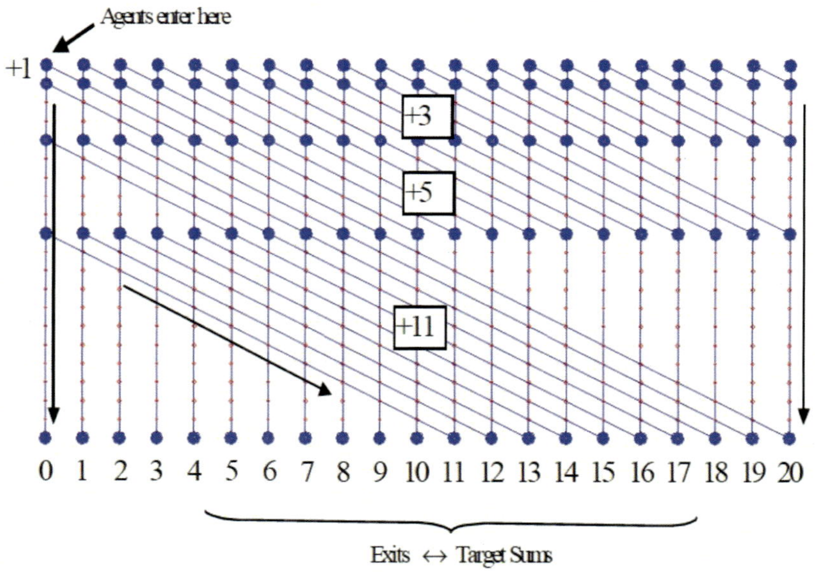

FIGURE 2
Schema for a digraph encoding for the problem SUBSET SUM. The input node, at the top left, is populated with agents at the beginning of the computation while all other nodes are unpopulated. The node Schematic of general computational process on N-systems. Agents start at the input node(s) and percolate stochastically through the encoding of a problem instance by the

A solution $\Sigma w_i s_i = T$ to the SSP is represented by that subset of agents choosing a path to one of the output nodes in the network (bottom row in Fig. 2). Since we here consider the special case of an unlimited number of agents, this is guaranteed to lead to the state $V_\Omega(t) = 1$ for all exit nodes Ω corresponding to a legal sum over some subset of S, since all possible paths are explored, and therefore all possible subset sums of S are generated, simultaneously.

Since all NP-complete problems can be converted to one another in polynomial time, this particular digraph encoding (for SUBSET SUM) can, in theory, be used to solve any other NP-complete problem, by conversion to a SUBSET SUM instance. However, the existence of these conversions, guaranteed by the Cook-Levin Theorem [47] does not imply that the resulting encoding into SUBSET SUM form from the original problem is "compact" (polynomially-sized in the size of the original instance). For example, the known reduction from 3-SAT to SUBSET SUM [14] would result in a very large SUBSET SUM network. Therefore, it is desirable to find compact encodings, using efficient reductions, for individual NP-complete problems.

5 THE COMPUTING POWER OF N-SYSTEMS

"Computing power" and similar expressions have (at least) two broad meanings. The first relates to what kinds of functions a system is able to compute given enough/unbounded resources (time, space, energy, etc.), in the sense in which Turing was interested was in understanding computational machines. Thus, systems that are able to compute all Turing-computable functions are said to be *Turing complete* or *computationally universal*. Intuitively, these systems can compute all functions in which we would be interested from a practical point of view. Important examples of computationally complete systems are, clasically, the lambda calculus [10] and cellular automata [67], while bio-inspired paradigms include P-systems [53], recurrent neural networks [44] and DNA [12] and quantum computing [18].

This concept of computational completeness tells us about what "access" a system may have to different classes of function, but says nothing about the "efficiency" of the computational processes. Computing "power" is also sometimes used to refer to the use of resources scales for a given computing paradigm for different functions and, importantly, how these resources scale with problem instance size. Thus, if P!=NP, as is generally believed [1], then classical, sequential computers cannot, in the worst case, realistically solve NP-complete problem instances beyond a modest instance size. Although factorisation is not known (or thought) to be NP-complete, it is a function believed to be "hard" to compute with sequential machines and this property is the basis of factorisation-based cryptography. Here, "hard" is by convention taken to mean that, as the number of bits B required to specify the instance size grows, either the time or the space required by an "algorithm" running on a given system (in the worst case) grows superpolynomially with B, while, in contrast, "easy" is taken to mean that the relevant resource grows as a polynomial in B.

A simple observation based on the example in the last section, SUBSET SUM, is that we can design a simple N-system that can enumerate the integers from 0 to k, for example, by adapting our SUBSET sum network so that the set of integers is replaced by the multiset of cardinality k+1, namely, {0, 1, 1 … }, i.e. with k occurrences of 1. Seeding such a system with agents at the input junction representing a sum of 0 and prior to any branches (as above), we get the N-system illustrated in Fig 3 below. If we consider all positive integers smaller than k to be target sums in the SUBSET SUM sense, we obtain, given enough agents and steps, all integers between from 0 to k (k=10 in Figure 3).

We now turn to issues regarding the computing power of N-systems in the specific contexts of combinatorial problems and resource requirements, respectively. We are interested, at a high level, in what improvement (if any), N-system may offer over sequential, classical computers.

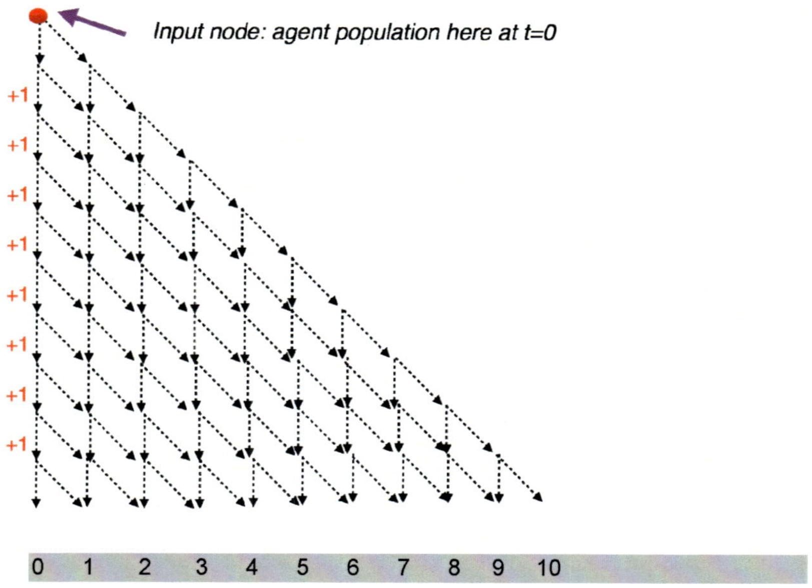

FIGURE 3
Digraph encoding for SUBSET SUM on the multiset {1, 1, 1 …}. An agent population initially at the input node (with all other nodes unvisited) percolates through this encoding such that, after 10 computational steps, all output nodes are visited, i.e. this device performs an enumeration of the first 11 integers.

We begin with two observations. The first is that the *Agent_Paths* program illustrated using pseudocode in Fig 4 simulates on a Turing machine the operation of an N-system specified by a digraph G and an initial "large" population of n agents, starting at locations node(0,1:n). Note that, while an N-system would operate in parallel, i.e. all agents would move simultaneously at each step, while the (sequential) Turing machine is only able to move each of them one at a time. Therefore, if the running time of the N-system might grow as $O(T)$, the program would have a running time of $O(nT)$ steps. We will rely on the observation that this programme simulates an N-system in establishing the results below.

Secondly, putting aside running time, computational efficiency and the specific problem and instance an N-system is computing, we see that N-systems are actually algorithms for stochastically determining connectivity between two nodes (the input node and the output node) in a given digraph G. More specifically, this problem, known as st-connectivity (STCON), is defined as a decision problem:

$$\text{PATH} = \{\langle G, s, t\rangle \mid G \text{ is a directed graph}\} \tag{5}$$

In the case where G has a path between s and t, given a sufficiently large number of agents and sufficiently many steps, it is clear that the N-system (and Agent_Paths) will find it, in the following sense: as the number of agents becomes larger than the number of total paths through the graph, the probability that, given an arbitrarily large number of steps t_max, an agent will not reach the output node tends to 0, irrespective of the connectivity pattern of G.

In complexity theory terms, STCON is actually NL-complete, that is, every problem in the class NL is reducible to connectivity under a log-space reduction [29]. The logspace reduction from any language in NL to STCON proceeds as follows: Consider the nondeterministic log-space Turing machine M that accepts a language in *NL*. Since there is only logarithmic space on the work tape, all possible states of the Turing machine (where a state is the state of the internal finite state machine, the position of the head and the contents of the work tape) are polynomially many. Map all possible states of the deterministic log-space machine to vertices of a graph, and put an edge between u and v if the state v can be reached from u within one step of the non-deterministic machine. Now the problem of whether the machine accepts is the same as the problem of whether there exists a path from the start state to the accepting state.

It is known that NL is contained in P, since there is a polynomial-time algorithm for 2-satisfiability (Aaronson, 1996). It is not known whether NL = P (Aaronson, 1996). It is also, of course, not known whether P != NP. We therefore have the relation NL $\subseteq$ P $\subseteq$ NP.

With the observation that N-systems with arbitrarily large numbers of agents are parallel inplementations of the program Agent_Paths, we are now ready for the first of three results regarding the computing power of N-systems.

Proposition 1. If a digraph encoding of an NP-complete problem D exists that is polynomial in the number $|D_X|$ of bits required to describe problem instance D_X, and that can be found in time polynomial in $|D_X|$, then P = NP.

Sketch of proof. Suppose that for any instance D_X of D, there exists an encoding $D_X \rightarrow E_X$ that can be generated in time polynomial in $|D_X|$. If such an E_X has size $| E_X|$ (number of nodes) polynomial in $| D_X|$ then the problem of finding a solution to D_X reduces to that of determining the existence of a path on E_X that connects the input node I and output node Ω on E_X, i.e. solving the STCONN problem instance

$$Path\_E_X = \{\langle E_X, I, \Omega\rangle \mid E_X \text{ is a directed graph with a path from vertex I to } \Omega\}.$$

STCONN is in the class NL $\subseteq$ P, i.e. in a member of P. This would imply that $Path\_E_X$ and therefore D_X can be solved in polynomial time on a Turing machine. Then P = NP.

The informal interpretation of this result is simple. An encoding E_X can always be found that is polynomial in $|D_X|$, but, if P != NP, this cannot be done in polynomial time on a Turing machine since it would essentially be equivalent to directly solving the problem instance. The implication for N-system of the type described here, even those possessing an arbitrarily large number of agents, is that if P != NP, the encodings, in general, are not polynomial in size (have a number of nodes polynomial in $|D_X|$). This leads directly to our second result.

Proposition 2. If P != NP, N-systems with arbitrarily many agents cannot solve NP-complete problems in polynomial time.

Sketch of proof. Consider an NP-complete problem D and suppose that for any instance D_X of D, there exists an encoding $D_X \rightarrow E_X$. Observe that the running time of any N-system K can be written as $T = T_{enconding} + T_K + C$, where $T_{enconding}$ is the time required to find E_X (by any Turing machine), T_K is the running time of K on E_X and C is a constant. Suppose P != NP and there exists an N-system K that solves any instance D_X of D in polynomial time. Then T is a polynomial in $|D_X|$. By Proposition 1, if $T_{enconding}$ is polynomial in $|D_X|$ and $|E_X|$ is polynomial in $|D_X|$, then P = NP, which contradicts our assumption that P != NP. Therefore, for $T_{enconding}$ polynomial (or smaller) in $|D_X|$, there does not exist an integer q such that $|E_X| = |D_X|^q + |D_X|$, i.e. $|E_X|$ is not polynomial. Now observe that the running time of the program *Agent_paths* above is $O(m^2)$, i.e. $O(|E_X|^2)$. The N-system K is simply executing the inner loop of *Agent_paths* in parallel (all agents move at once at each computing step). Its running time, T_K, is therefore $O(T_{Agent\_paths} / |E_X|) = O(|E_X|)$. Since $|E_X|$ is not polynomial, T_K is therefore also not polynomial.

```
Program Agent_Paths

input: W(n,n), V_I, V_Ω(0) // the digraph encod-
ing (m nodes) and the input/output nodes; out-
put: 1 iff V_Ω(t) = 1 and t < t_max, else 0
t = 0;
V_I(0) = 1; V_i(0) = 0 for all other i.
while t < t_max and V_Ω(t) = 0
     for i = 1 to m
          H_i = 0
          for j = 1 to m
               H_i = H_i + W(j,i)*V_j(t)
          end
          V_i(t) = sgn(H_i)
     t = t + 1 end
```

FIGURE 4
Pseudocode for the program Agent_Paths.

This result can be thought of informally in this way: N-system as described above are massively parallel devices for solving STCONN once furnished with an encoding. They achieve this goal through a breadth-first-like stochastic search algorithm in which all agents that are able to move, do so simultaneously at each time step, effectively executing the innermost loop of *Agent_paths* in a constant time step. Their "total" running time is the time to find the encoding for a problem, by any means (e.g. electronic computer) plus their own running time to solve STCONN in parallel. If P != NP, either the encoding of any problem is polynomial, requiring superpolynomial time to find (in which case the N-systemhas total running time that is superpolynomial); or the encoding can be found in polynomial time but cannot itself be polynomial, in which case the N-system itself cannot have polynomial running time.

6 DISCUSSION

This paper describes a new model of computation, of the distributed parallel type, based on the directed motions of autonomous physical agents exploring physical realisations of specially designed networks. In this model, the paths of the agents through the network represent partial "running computations" while the visitation status of the network's nodes either in the stable state or at some prescribed time represents the result of the computation. We call the resulting computing network devices "*N-systems*" and we investigate the theoretical computational power of a simplified subset of N-systems, respectively those that in the initial state are "loaded" with an unlimited number of agents, such that we do not need to be concerned about the population of agents becoming diluted as it explores the network stochastically.

While the two results in the section above appear negative regarding the computational power of N-systems, it is easy to see, on the other hand, that the agents are in fact explicity exploring every path through the network in parallel (as well as solving connectivity). Therefore, if we had, in addition to the above properties, the ability for individual agents to "know" their paths through the network and somehow send that information "upwards", then these devices could conceivably solve NP-complete problems in polynomial time.

To see that this is indeed the case, consider any instance of the directed Travelling Salesman Problem (TSP) with n nodes and a population of agents located initially at any node p_1. By definition, if the graph contains a self-avoiding loop $p_1 p_2 p_3 \dots p_1$ of length n+1, i.e. starting and ending at p_1, then, assuming the initial existence of enough agents at p_1, after n+1 steps, that path will have been traversed by one (or many) agents. If, having completed this path, the relevant agent or agents can output their trajectory (or merely its existence), then the system can solve the TPS in linear time (n+1 steps). We call versions of N-systems in which the agents can report their own path information during runtime "sticker" N-systems. While these are beyond the scope of this paper, given their clear potential to solve combinatorial problems efficiently, the more formal investigation of their computing power will will form the basis of future work.

In the physical implementation of N-systems, it is conceivable to use agents that are able to self-replicate (e.g. bacteria) as they travel through the encoding network. In fact, this is desirable, since the N-system is essentially stochastically visiting all the paths from input to output and, therefore, "organically" increasing the number of agents would directly amplify the computing power of a given N-system and allow it to naturally scale with problem instance size. This would be especially true if when an agent replicates or otherwise produces a progeny, the daughter agent would have a record of the trajectory of the parent through the network up to the point of replication. Investigations of the properties of systems equipped with such agents will be the subject of future work.

In practice, physical realisations of N-systems, such as that demonstrated in [49], are equipped with a potentially very large but not unlimited number of agents. Conceptually, this is analogous to DNA computing devices [5], which are able to form potential solutions through a problem by combinations of DNA particles in solution. These devices can use astronomically large numbers of DNA molecules, but the number is finite. Indeed, due to the presence of errors in the solution-generation and amplification steps, this physical limitation is one of the reasons why DNA computers have (so far), not been able to scale enough to solve realistic-size NP-complete problems or to challenge electronic computers generally.

Therefore, studying more realistic N-system models, i.e. those in which the initial population of agents is finite, and therefore in which the population of agents at any node throughout the network is dynamic during the computation (and finite), is a necessary step in the further development of this model of computation. Furthermore, and as shown even in our first realisation of an N-system (Nicolau et al., 2016), in practice, physical agents exploring networks comprised of nonzero-width channels occasionally make navigation errors, and this can be thought of as a type of error in the local computational steps of the N-system. Thirdly, since physical agents have nonzero size, crowding and/or collisions between them are issues that must be taken into account. Finally, as suggested by the TSP example above, "sticker" N-systems would be dramatically more powerful as computing devices. Investigations of the theoretical properties of these more realistic models of physical N-systemsare the subject of future work.

Quantum computing has received a great deal of attention both theoretically and experimentally [8]. Since both models operate in a massively parallel way, it is interesting to speculate, even at this early stage, ton their respective theoretical abilities to solve NP-complete problems specifically. We note that sticker N-systems, can clearly in theory solve NP-complete problems in polynomial time — as described above regarding TSP and also by noting their similarity to DNA computers, which are known to have this property. In contrast, we note that the best theoretical results known at present indicate that quantum computers, even in their most ideal implementation, would at best offer a square-root improvement over electronic computers and therefore are currently believed to be unable to solve NP-complete problems efficiently (in the sense of requiring polynomial time).

Finally, there are some clear similarities between N-systemsand theoretical models of neural networks (in the computer science and physiological sense, respectively). More realistic models of neural circuits, such as spiking neural nets (SNNs) [68], have the nodes communicate with each other by sending individual pulses. N-system of the type described in this section, with finite agent numbers and therefore discrete, individual agents passing along arcs in the digraph encodings between nodes, share this feature at a high level. This observation suggests an open problem: are models like SNNs and N-systems computationally equivalent? Little is known formally about the computational power of SNNs [68], but recurrent neural networks (RNNs), a simplified model of SNNs, are known to be computationally complete in the sense of being able to simulate any Turing machine [58]. In this context, comparative studies of the computational efficiency and formal power of N-systems and SNNs/RNNs shall form the subject of future work.

7 CONCLUSION

This paper formally introduces a new model of computation, of the distributed parallel type, based on the exploration of specially designed digraph encodings of problems by populations of autonomous, mobile agents transitioning between nodes. We studied the potential computing power and limitations of such a system, which we call N-systems, to solve combinatorial problems efficiently, showing that, in principle, the model is capable of solving NP-complete problems in polynomial time (even if P!=NP), but only if equipped with very large numbers of agents relative to the problem size and if the agents are able to retain some information about their trajectories.

ACKNOWLEDGEMENTS

The author gratefully acknowledges support for this work from the European Union's Bio4Comp Consortium under the auspices of the EU's Horizon2020 framework and Australian Research Council Future Fellowship FT180100698. I would like to thank Dan V. Nicolau, Sr. for fruitful discussions that have improved this manuscript.

REFERENCES

[1] Aaronson, Scott. '$$P\mathop{ =}\limits^{?}NP$$'. *Open Problems in Mathematics*, edited by Jr. Nash John Forbes and Michael Th. Rassias, Springer International Publishing, 2016, pp. 1–122. *Springer Link*, https://doi.org/10.1007/978-3-319-32162-2_1.

[2] Abe, Kaori, et al. 'Simultaneous Immunoassay Analysis of Plasma IL-6 and TNF- on a Microchip'. *PLOS ONE*, vol. 8, no. 1, Jan. 2013, p. e53620. *PLoS Journals*, https://doi.org/10.1371/journal.pone.0053620.

[3] Ader, R., and N. Cohen. 'Behaviorally Conditioned Immunosuppression'. *Psychosomatic Medicine*, vol. 37, no. 4, 1975, pp. 333–40. *PubMed*, https://doi.org/10.1097/00006842-197507000-00007.

[4] Ader, Robert, and Nicholas Cohen. 'Behaviorally Conditioned Immunosuppression'. *Psychosomatic Medicine*, vol. 37, no. 4, Aug. 1975, p. 333. *journals.lww.com*, https://journals.lww.com/psychosomaticmedicine/abstract/1975/07000/behaviorally_conditioned_immunosuppression.7.aspx.

[5] Adleman, Leonard M. 'Computing with DNA'. *Scientific American*, vol. 279, no. 2, 1998, pp. 54–61. *JSTOR*, https://www.jstor.org/stable/26070598.

[6] Alzubaidi, Laith, et al. 'Review of Deep Learning: Concepts, CNN Architectures, Challenges, Applications, Future Directions'. *Journal of Big Data*, vol. 8, no. 1, Mar. 2021, p. 53. *BioMed Central*, https://doi.org/10.1186/s40537-021-00444-8.

[7] Amemiya, Takashi, et al. 'Glycolytic Oscillations in HeLa Cervical Cancer Cell Spheroids'. *The FEBS Journal*, vol. 289, no. 18, Sept. 2022, pp. 5551–70. *DOI.org (Crossref)*, https://doi.org/10.1111/febs.16454.

[8] Arute, Frank, et al. 'Quantum Supremacy Using a Programmable Superconducting Processor'. *Nature*, vol. 574, no. 7779, Oct. 2019, pp. 505–10. *www.nature.com*, https://doi.org/10.1038/s41586-019-1666-5.

[9] Bai, Fan, et al. 'Conformational Spread as a Mechanism for Cooperativity in the Bacterial Flagellar Switch'. *Science (New York, N.Y.)*, vol. 327, no. 5966, Feb. 2010, pp. 685–89. *PubMed*, https://doi.org/10.1126/science.1182105.

[10] Barendregt, Henk. 'The Impact of the Lambda Calculus in Logic and Computer Science'. *The Bulletin of Symbolic Logic*, vol. 3, no. 2, 1997, pp. 181–215. *JSTOR*, https://doi.org/10.2307/421013.

[11] Battle, Alexis, et al. 'Impact of Regulatory Variation from RNA to Protein'. *Science*, vol. 347, no. 6222, Feb. 2015, pp. 664–67. *DOI.org (Crossref)*, https://doi.org/10.1126/science.1260793.

[12] Boneh, Dan, et al. 'On the Computational Power of DNA'. *Discrete Applied Mathematics*, vol. 71, no. 1, Dec. 1996, pp. 79–94. *ScienceDirect*, https://doi.org/10.1016/S0166-218X(96)00058-3.

[13] Caprara, Alberto, et al. 'The Multiple Subset Sum Problem'. *SIAM Journal on Optimization*, vol. 11, no. 2, Jan. 2000, pp. 308–19. *DOI.org (Crossref)*, https://doi.org/10.1137/S1052623498348481.

[14] Cardinal, Jean, et al. 'Reconfiguration of Satisfying Assignments and Subset Sums: Easy to Find, Hard to Connect'. *Theoretical Computer Science*, vol. 806, Feb. 2020, pp. 332–43. *ScienceDirect*, https://doi.org/10.1016/j.tcs.2019.05.028.

[15] Cass, Steven P., et al. 'Moving the Pathway Goalposts: COPD as an Immune-Mediated Inflammatory Disease'. *The Lancet Respiratory Medicine*, vol. 10, no. 12, 2022, pp. 1110–13.

[16] Chiu, Daniel T., et al. 'Using Three-Dimensional Microfluidic Networks for Solving Computationally Hard Problems'. *Proceedings of the National Academy of Sciences*, vol. 98, no. 6, Mar. 2001, pp. 2961–66. *DOI.org (Crossref)*, https://doi.org/10.1073/pnas.061014198.

[17] Chung, Hae Young, et al. 'Molecular Inflammation: Underpinnings of Aging and Age-Related Diseases'. *Ageing Research Reviews*, vol. 8, no. 1, Jan. 2009, pp. 18–30. *PubMed*, https://doi.org/10.1016/j.arr.2008.07.002.

[18] Clément, Alexandre, et al. 'Quantum Circuit Completeness: Extensions and Simplifications'. *arXiv.Org*, 6 Mar. 2023, https://arxiv.org/abs/2303.03117v3.

[19] Colin, Remy, et al. 'Multiple Sources of Slow Activity Fluctuations in a Bacterial Chemosensory Network'. *eLife*, edited by Naama Barkai, vol. 6, Dec. 2017, p. e26796. *eLife*, https://doi.org/10.7554/eLife.26796.

[20] Eberl, Gerard. 'A New Age for (Mucosal) NeuroImmunology'. *Mucosal Immunology*, vol. 15, no. 6, June 2022, pp. 1052–55. *www.nature.com*, https://doi.org/10.1038/s41385-022-00573-0.

[21] Eldar, Avigdor, and Michael B. Elowitz. 'Functional Roles for Noise in Genetic Circuits'. *Nature*, vol. 467, no. 7312, Sept. 2010, pp. 167–73. *DOI.org (Crossref)*, https://doi.org/10.1038/nature09326.

[22] Elowitz, Michael B., et al. 'Stochastic Gene Expression in a Single Cell'. *Science*, vol. 297, no. 5584, Aug. 2002, pp. 1183–86. *DOI.org (Crossref)*, https://doi.org/10.1126/science.1070919.

[23] Flajnik, Martin F., and Masanori Kasahara. 'Origin and Evolution of the Adaptive Immune System: Genetic Events and Selective Pressures'. *Nature Reviews. Genetics*, vol. 11, no. 1, Jan. 2010, pp. 47–59. *PubMed Central*, https://doi.org/10.1038/nrg2703.

[24] Franceschi, Claudio, and Judith Campisi. 'Chronic Inflammation (Inflammaging) and Its Potential Contribution to Age-Associated Diseases'. *The Journals of Gerontology. Series A, Biological Sciences and Medical Sciences*, vol. 69 Suppl 1, June 2014, pp. S4-9. *PubMed*, https://doi.org/10.1093/gerona/glu057.

[25] Garey, Michael R. *Computers and Intractability : A Guide to the Theory of NP-Completeness*. San Francisco : W. H. Freeman, 1979. *Internet Archive*, http://archive.org/details/computersintract0000gare.

[26] Gruol, Donna L. 'The Neuroimmune System and the Cerebellum'. *The Cerebellum*, Nov. 2023. *Springer Link*, https://doi.org/10.1007/s12311-023-01624-3.

[27] Gyongyosi, Laszlo, and Sandor Imre. 'A Survey on Quantum Computing Technology'. *Computer Science Review*, vol. 31, Feb. 2019, pp. 51–71. *ScienceDirect*, https://doi.org/10.1016/j.cosrev.2018.11.002.

[28] Hodes, Georgia E., et al. 'Neuroimmune Mechanisms of Depression'. *Nature Neuroscience*, vol. 18, no. 10, Oct. 2015, pp. 1386–93. *PubMed Central*, https://doi.org/10.1038/nn.4113.

[29] Immerman, Neil. *Descriptive Complexity*. Springer Science & Business Media, 1998.

[30] Jansen, Joanneke E., et al. 'A Method for the Inference of Cytokine Interaction Networks'. *PLoS Computational Biology*, vol. 18, no. 6, June 2022, p. e1010112. *PubMed Central*, https://doi.org/10.1371/journal.pcbi.1010112.

[31] Jin, Hao, et al. 'A Body–Brain Circuit That Regulates Body Inflammatory Responses'. *Nature*, May 2024, pp. 1–9. *www.nature.com*, https://doi.org/10.1038/s41586-024-07469-y.

[32] Kamimura, Daisuke, et al. 'Bidirectional Communication between Neural and Immune Systems'. *International Immunology*, vol. 32, no. 11, Oct. 2020, pp. 693–701. *PubMed*, https://doi.org/10.1093/intimm/dxz083.

[33] Kate, Aniket, and Ian Goldberg. 'Generalizing Cryptosystems Based on the Subset Sum Problem'. *International Journal of Information Security*, vol. 10, no. 3, June 2011, pp. 189–99. *Springer Link*, https://doi.org/10.1007/s10207-011-0129-2.

[34] Keegstra, Johannes M., et al. 'Phenotypic Diversity and Temporal Variability in a Bacterial Signaling Network Revealed by Single-Cell FRET'. *eLife*, edited by Naama Barkai, vol. 6, Dec. 2017, p. e27455. *eLife*, https://doi.org/10.7554/eLife.27455.

[35] Kellerer, Hans, et al. 'An Efficient Fully Polynomial Approximation Scheme for the Subset-Sum Problem'. *Journal of Computer and System Sciences*, vol. 66, no. 2, Mar. 2003, pp. 349–70. *ScienceDirect*, https://doi.org/10.1016/S0022-0000(03)00006-0.

[36] Koren, Tamar, et al. 'Insular Cortex Neurons Encode and Retrieve Specific Immune Responses'. *Cell*, vol. 184, no. 24, Nov. 2021, pp. 5902-5915.e17. *ScienceDirect*, https://doi.org/10.1016/j.cell.2021.10.013.

[37] ---. 'Insular Cortex Neurons Encode and Retrieve Specific Immune Responses'. *Cell*, vol. 184, no. 24, Nov. 2021, pp. 5902-5915.e17. *PubMed*, https://doi.org/10.1016/j.cell.2021.10.013.

[38] Kraus, Aurora, et al. 'Sensing the World and Its Dangers: An Evolutionary Perspective in Neuroimmunology'. *eLife*, vol. 10, Apr. 2021, p. e66706. *PubMed*, https://doi.org/10.7554/eLife.66706.

[39] Kubiczkova, Lenka, et al. 'TGF-β – an Excellent Servant but a Bad Master'. *Journal of Translational Medicine*, vol. 10, no. 1, Sept. 2012, p. 183. *BioMed Central*, https://doi.org/10.1186/1479-5876-10-183.

[40] Levine, Erel, and Terence Hwa. 'Stochastic Fluctuations in Metabolic Pathways'. *Proceedings of the National Academy of Sciences*, vol. 104, no. 22, May 2007, pp. 9224–29. *DOI.org (Crossref)*, https://doi.org/10.1073/pnas.0610987104.

[41] Li, Suny. 'From Moore's Law to Function Density Law'. *MicroSystem Based on SiP Technology*, edited by Suny Li, Springer Nature, 2022, pp. 3–27. *Springer Link*, https://doi.org/10.1007/978-981-19-0083-9_1.

[42] Livneh, Yoav, et al. 'Estimation of Current and Future Physiological States in Insular Cortex'. *Neuron*, vol. 105, no. 6, Mar. 2020, pp. 1094-1111.e10. *PubMed*, https://doi.org/10.1016/j.neuron.2019.12.027.

[43] ---. 'Estimation of Current and Future Physiological States in Insular Cortex'. *Neuron*, vol. 105, no. 6, Mar. 2020, pp. 1094-1111.e10. *PubMed*, https://doi.org/10.1016/j.neuron.2019.12.027.

[44] Mali, Ankur, et al. 'On the Computational Complexity and Formal Hierarchy of Second Order Recurrent Neural Networks'. *arXiv.Org*, 26 Sept. 2023, https://arxiv.org/abs/2309.14691v1.

[45] Meade, Elaine, and Mary Garvey. 'The Role of Neuro-Immune Interaction in Chronic Pain Conditions; Functional Somatic Syndrome, Neurogenic Inflammation, and Peripheral Neuropathy'. *International Journal of Molecular Sciences*, vol. 23, no. 15, Aug. 2022, p. 8574. *PubMed Central*, https://doi.org/10.3390/ijms23158574.

[46] Merrins, Matthew J., et al. 'Direct Measurements of Oscillatory Glycolysis in Pancreatic Islet β-Cells Using Novel Fluorescence Resonance Energy Transfer (FRET) Biosensors for Pyruvate Kinase M2 Activity'. *The Journal of Biological Chemistry*, vol. 288, no. 46, Nov. 2013, pp. 33312–22. *PubMed*, https://doi.org/10.1074/jbc.M113.508127.

[47] Miller, Raymond E., et al., editors. *Complexity of Computer Computations: Proceedings*. Plenum Press, 1972.

[48] Mojica-Benavides, Martin, et al. 'Intercellular Communication Induces Glycolytic Synchronization Waves between Individually Oscillating Cells'. *Proceedings of the National Academy of Sciences*, vol. 118, no. 6, Feb. 2021, p. e2010075118. *DOI.org (Crossref)*, https://doi.org/10.1073/pnas.2010075118.

[49] Nicolau, Dan V., et al. 'Molecular Motors-Based Micro- and Nano-Biocomputation Devices'. *Microelectronic Engineering*, vol. 83, no. 4, Apr. 2006, pp. 1582–88. *ScienceDirect*, https://doi.org/10.1016/j.mee.2006.01.198.

[50] Ottaviani, Enzo, et al. 'Common Evolutionary Origin of the Immune and Neuroendocrine Systems: From Morphological and Functional Evidence to *in Silico* Approaches'. *Trends in Immunology*, vol. 28, no. 11, Nov. 2007, pp. 497–502. *ScienceDirect*, https://doi.org/10.1016/j.it.2007.08.007.

[51] Papagiannakis, Alexandros, et al. 'Autonomous Metabolic Oscillations Robustly Gate the Early and Late Cell Cycle'. *Molecular Cell*, vol. 65, no. 2, Jan. 2017, pp. 285–95. *PubMed*, https://doi.org/10.1016/j.molcel.2016.11.018.

[52] Patel, Pragna, et al. 'Clinical Characteristics of Multisystem Inflammatory Syndrome in Adults: A Systematic Review'. *JAMA Network Open*, vol. 4, no. 9, Sept. 2021, p. e2126456. *Silverchair*, https://doi.org/10.1001/jamanetworkopen.2021.26456.

[53] Păun, Gheorghe. *Computing with Membranes*. Turku Centre for Computer Science, 1998.

[54] Powell, James R. 'The Quantum Limit to Moore's Law'. *Proceedings of the IEEE*, vol. 96, no. 8, Aug. 2008, pp. 1247–48. *IEEE Xplore*, https://doi.org/10.1109/JPROC.2008.925411.

[55] Rea, Irene Maeve, et al. 'Age and Age-Related Diseases: Role of Inflammation Triggers and Cytokines'. *Frontiers in Immunology*, vol. 9, 2018. *Frontiers*, https://www.frontiersin.org/articles/10.3389/fimmu.2018.00586.

[56] Scheff, Nicole N., and Jami L. Saloman. 'Neuroimmunology of Cancer and Associated Symptomology'. *Immunology and Cell Biology*, vol. 99, no. 9, Oct. 2021, pp. 949–61. *PubMed Central*, https://doi.org/10.1111/imcb.12496.

[57] Schnorr, C. P., and M. Euchner. 'Lattice Basis Reduction: Improved Practical Algorithms and Solving Subset Sum Problems'. *Mathematical Programming*, vol. 66, no. 1, Aug. 1994, pp. 181–99. *Springer Link*, https://doi.org/10.1007/BF01581144.

[58] Siegelmann, H. T., and E. D. Sontag. 'On the Computational Power of Neural Nets'. *Journal of Computer and System Sciences*, vol. 50, no. 1, Feb. 1995, pp. 132–50. *ScienceDirect*, https://doi.org/10.1006/jcss.1995.1013.

[59] Sosík, Petr, et al. 'From P Systems to Morphogenetic Systems: An Overview and Open Problems'. *Journal of Membrane Computing*, vol. 2, no. 4, Dec. 2020, pp. 380–91. *Springer Link*, https://doi.org/10.1007/s41965-020-00057-9.

[60] Sugimoto, Michelle A., et al. 'Resolution of Inflammation: What Controls Its Onset?' *Frontiers in Immunology*, vol. 7, 2016. *Frontiers*, https://www.frontiersin.org/articles/10.3389/fimmu.2016.00160.

[61] Tan, Li Yin, et al. 'Hyperinflammatory Immune Response and COVID-19: A Double Edged Sword'. *Frontiers in Immunology*, vol. 12, 2021. *Frontiers*, https://www.frontiersin.org/articles/10.3389/fimmu.2021.742941.

[62] ter Horst, Rob, et al. 'Host and Environmental Factors Influencing Individual Human Cytokine Responses'. *Cell*, vol. 167, no. 4, Nov. 2016, pp. 1111-1124.e13. *ScienceDirect*, https://doi.org/10.1016/j.cell.2016.10.018.

[63] Theis, Thomas N., and H. S. Philip Wong. 'The End of Moore's Law: A New Beginning for Information Technology'. *Computing in Science & Engineering*, vol. 19, no. 2, Mar. 2017, pp. 41–50. *IEEE Xplore*, https://doi.org/10.1109/MCSE.2017.29.

[64] Van Delft, Falco C. M. J. M., et al. 'Something Has to Give: Scaling Combinatorial Computing by Biological Agents Exploring Physical Networks Encoding NP-Complete Problems'. *Interface Focus*, vol. 8, no. 6, Dec. 2018, p. 20180034. *DOI.org (Crossref)*, https://doi.org/10.1098/rsfs.2018.0034.

[65] Wang, Fei, et al. 'Implementing Digital Computing with DNA-Based Switching Circuits'. *Nature Communications*, vol. 11, no. 1, Jan. 2020, p. 121. *www.nature.com*, https://doi.org/10.1038/s41467-019-13980-y.

[66] Williamson, Thomas, et al. 'Exploring the Genetic Control of Glycolytic Oscillations in Saccharomyces Cerevisiae'. *BMC Systems Biology*, vol. 6, no. 1, Aug. 2012, p. 108. *BioMed Central*, https://doi.org/10.1186/1752-0509-6-108.

[67] Yamashita, Tatsuya, et al. 'Turing-Completeness of Asynchronous Non-Camouflage Cellular Automata'. *Information and Computation*, vol. 274, Oct. 2020, p. 104539. *ScienceDirect*, https://doi.org/10.1016/j.ic.2020.104539.

[68] Yamazaki, Kashu, et al. 'Spiking Neural Networks and Their Applications: A Review'. *Brain Sciences*, vol. 12, no. 7, July 2022, p. 863. *www.mdpi.com*, https://doi.org/10.3390/brainsci12070863.

[69] Zarate, Sara M., et al. 'Neuroimmunology of Cardiovascular Disease'. *Current Hypertension Reports*, Apr. 2024. *Springer Link*, https://doi.org/10.1007/s11906-024-01301-8.

[70] Zhang, Zheng, et al. 'Dynamic Single-Cell NAD(P)H Measurement Reveals Oscillatory Metabolism throughout the E. Coli Cell Division Cycle'. *Scientific Reports*, vol. 8, no. 1, Feb. 2018, p. 2162. *www.nature.com*, https://doi.org/10.1038/s41598-018-20550-7.

[71] ---. 'Dynamic Single-Cell NAD(P)H Measurement Reveals Oscillatory Metabolism throughout the E. Coli Cell Division Cycle'. *Scientific Reports*, vol. 8, no. 1, Feb. 2018, p. 2162. *PubMed*, https://doi.org/10.1038/s41598-018-20550-7.

Comparison of Effectiveness: Biological vs AI Systems

JORDI VALLVERDÚ[1] AND MAX TALANOV[2,3,4,*]

[1]*ICREA Acadèmia - UAB, Barcelona, Catalonia, Spain*
[2]*Department of Engineering, University of Messina, Messina, Italy*
[3]*Institute for Artificial Intelligence R&D of Serbia, Novi Sad, Serbia*
[4]*ITIS, Kazan Federal University, Kazan, Russian Federation*

Received: September 1, 2024. Accepted: September 15, 2024.

In this work, we present a comparison of the effectiveness of biological systems and modern digital AI solutions. Life, from its simplest forms such as slime molds to complex biological systems with central nervous systems, operates as a physical process counteracting entropy. This paper explores the relationship between energy and life, focusing on how living organisms minimise and optimise energy consumption to maintain their highly ordered states. Additionally, it compares the effectiveness of biological systems in energy optimization with digital artificial intelligence (AI) systems. By analysing mechanisms across various scales of biological complexity, this paper highlights the principles of negentropy in living organisms and evaluates how these principles are totally not used in modern AI systems making them ineffective and outdated.

Keywords: Energy consumption, ANN, SNN, AI

1 INTRODUCTION

The second law of thermodynamics states that the entropy of an isolated system will always increase over time, leading to a state of maximum disorder [22]. However, life exists as an exception, continually maintaining and creating order [12]. This paradox is resolved by considering living organisms as open systems that exchange energy and matter with their surroundings.

* Contact: author: E-mail: max.talanov@gmail.com

Life harnesses energy to build and maintain complex structures, effectively acting as a negentropic system [12]. This paper aims to understand how biological systems achieve this and how these strategies compare to those used in artificial intelligence systems.

2 BIOLOGICAL SYSTEM PROPERTIES

Negentropy, or negative entropy, is the process by which systems reduce their entropy by importing order from their environment [14]. Living organisms are prime examples of negentropic systems. They absorb energy from their environment (through processes such as photosynthesis or consumption of organic matter) and use it to fuel metabolic processes that build and maintain ordered structures [13].

Simple organisms like slime molds (Physarum polycephalum) provide clear examples of energy optimisation [1]. Slime molds exhibit intelligent behavior in foraging and navigating their environment [26]. Studies have shown that these organisms can solve complex problems, such as finding the shortest path through a maze, by optimizing their energy expenditure [18]. This optimization minimizes unnecessary energy consumption, reflecting a fundamental principle of biological efficiency. There are two other important characteristics to be studied:

- *Foraging Behaviour*: Slime moulds extend pseudopodia to explore their surroundings, retracting them upon encountering unfavorable conditions. This dynamic adjustment allows them to minimise energy use while maximising nutrient intake [17].
- *Network Formation*: Slime moulds form efficient networks that distribute nutrients throughout their structure. These networks are highly adaptable, reorganising in response to changes in the environment to maintain optimal efficiency [1].

In more complex organisms, such as those with central nervous systems, the principles of energy optimisation and negentropy are even more pronounced. The brain is an energy-intensive organ, consuming approximately 20 % of the body's total energy [7]. Neuronal networks are optimised for efficient signal transmission, reducing the energy cost of communication [16]. This optimisation is evident in the architecture of the brain, where myelinated axons speed up signal transmission and reduce energy consumption [19]. Complex organisms have evolved intricate metabolic pathways that regulate energy use. These pathways ensure that energy production and consumption are balanced, maintaining homeostasis. Hormonal regulation, such as insulin and glucagon in glucose metabolism, exemplifies how organisms fine-tune their energy expenditure. At the same time, higher organisms exhibit adaptive behaviours that minimise energy expenditure. For instance, animals optimise their foraging strategies and migratory patterns to reduce energy costs. In humans, behavioural adaptations, such as the use of tools and technology, further exemplify energy optimisation [23].

The mechanisms underlying energy optimisation in biological systems can be broadly categorised into structural, functional, behavioural, and social strategies [9]. Biological structures, from the molecular to the macroscopic level, are designed to maximise efficiency. Examples include the streamlined shapes of fish and birds that reduce drag, and the fractal branching of blood vessels that optimise nutrient distribution. Metabolic pathways are finely tuned to optimise energy production and use. Enzyme efficiency, substrate channelling, and feedback regulation all contribute to minimising energy waste. Organisms adopt behaviours that enhance energy efficiency. For example, social insects like ants and bees optimise their collective foraging efforts, while migratory birds follow routes that maximise energy conservation.Social cooperation among living systems or species represents a crucial strategy for optimising energy use and enhancing survival. Through cooperative behaviours, organisms can collectively reduce individual energy expenditures and avoid redundant processes, thereby unifying efforts and resources for mutual benefit. This section explores how social cooperation contributes to energy efficiency in various biological contexts:

- **Social Insects**: Social insects such as ants and bees display highly organized foraging behaviors that exemplify collective foraging and resource sharing. Ants, for instance, lay down pheromone trails to guide colony members to food sources efficiently, reducing the energy each individual would otherwise expend searching independently. Similarly, bees communicate the location of nectar sources through the waggle dance, enabling other bees to exploit these resources without redundant searching, thereby ensuring effective energy use across the colony. These cooperative behaviors highlight the sophisticated mechanisms of resource optimization in social insects [4].
- **Mammalian Social Structures**: Many mammals, such as wolves and lions, hunt in packs to take down larger prey with less individual energy expenditure compared to solitary hunting, thereby increasing overall energy intake relative to energy spent. Additionally, species like meerkats and some primates engage in parental care and cooperative breeding, where individuals other than the parents help raise the young [15]. This shared responsibility reduces the energy burden on any single individual and promotes higher survival rates for the offspring.

- **Human Societies**:Humans have developed agricultural practices and community living arrangements that optimize energy use, allowing for more efficient food production through the collective cultivation of crops and raising of livestock. Shared infrastructure, such as irrigation systems and storage facilities, further minimizes redundant energy expenditures. Additionally, human social cooperation has driven technological innovations that enhance energy efficiency; for instance, communal transportation systems like buses and trains are more energy-efficient than individual car usage [24].
- **Mutualistic Relationships**: Mutualistic relationships between species, such as those between plants and mycorrhizal fungi, exemplify energy optimization through cooperation [6]. In this symbiosis, fungi provide plants with essential soil nutrients, while plants supply fungi with carbohydrates produced via photosynthesis, creating a reciprocal exchange that optimizes energy use for both organisms. Similarly, cleaner fish and host fish engage in a mutualistic relationship where cleaner fish remove parasites from larger host fish, gaining a food source and allowing the host fish to save energy that would otherwise be spent dealing with parasites.

As we have shown, social cooperation among living systems and species demonstrates a fundamental strategy for optimizing energy use and enhancing survival. From the highly organized foraging behaviors of social insects to the collective hunting strategies of mammals, and from the agricultural practices of human societies to mutualistic relationships between different species, cooperative behaviors significantly reduce individual energy expenditures and prevent redundancy. These diverse examples underscore the critical role of cooperation in promoting efficiency and sustainability in various biological contexts, highlighting the intricate and sophisticated mechanisms that organisms have evolved to thrive collectively.

3 COMPARISON WITH AI SYSTEMS

Artificial intelligence systems, particularly those designed for tasks such as data processing and pattern recognition, can be compared to biological systems in terms of energy optimization. While AI systems do not face the same metabolic constraints as biological entities, they are designed to maximise computational efficiency and minimize energy use.

1. **Algorithmic Efficiency**: AI systems do not optimize energy use through inefficient algorithms especially taking into account

backpropagation in the worst case we indicated 6-8 orders of magnitude ineffectiveness of the digital ANNs. Techniques such as neural network pruning and feedback loops are not used in modern ANN approaches [21,25]. Even hardware accelerators used in current ANNs still indicate the 6 orders of magnitude higher energy consumption for computational load compared to biological NNs.

2. **Resource Allocation**: modern AI systems, are not similar to biological organisms and use and allocate resources in terabytes due to the ineffectiveness of the learning approaches not using self-organisation that is basic for biological systems. For instance, digital ANNs require exhausting datasets describing all possible input vectors and data augmentation makes the situation worse by making digital ANNs process all the possible vectors to become more-or-less flexible for real-world data processing [5].
3. **Learning and Adaptation**: Machine learning algorithms, demonstrate their inability to adapt their strategies to minimise energy consumption while maximizing performance this could be the reason of ChatGPT training using energy compared to the energy consumed by a small country like Sweden [3, 20].

4 DIFFERENCES AND INEFFECTIVENESS OF CURRENT DIGITAL AI SYSTEMS

Despite the remarkable advancements in artificial intelligence, current digital AI systems remain significantly less effective compared to biological brains, being 6-8 orders of magnitude less efficient. Neurons in biological brains operate with incredible energy efficiency, with the human brain consuming about 20 watts of power while performing complex tasks such as perception, learning, and decision-making. A single neuron can transmit about 1000 action potentials per second using approximately 1.64×10^{-9} joules per action potential, resulting in an energy consumption of about 1.64×10^{-6} joules per second [2]. Each action potential consumes roughly 4×10^{9} ATP molecules, with each ATP molecule providing about 1×10^{-20} joules of energy [8, 11]. In contrast, modern AI systems, particularly those involving deep learning and large-scale neural networks, require substantial computational power and energy, with data centers consuming megawatts of electricity [10]. A basic computer task, such as a simple arithmetic operation, consumes approximately 10^{-12} joules per operation on a modern processor. For comparison, training a deep neural network can consume energy in the order of kilowatt-hours (kWh), with state-of-the-art models consuming upwards of

1 kWh, equivalent to 3.6×10^6 joules. A single neuron consumes approximately 1.64 μJ per second, while a basic processor performing simple tasks can consume up to 1 joule per second, indicating that even simple digital tasks are several orders of magnitude more energy-intensive than the activities of a single neuron. The brain employs a highly interconnected network of neurons capable of processing and transmitting information simultaneously and adaptively, with synaptic plasticity enabling learning and memory formation through efficient modulation of synaptic strengths. Biological systems can generalize from limited data and adapt to new situations with minimal retraining, while current AI models require vast amounts of labeled data for training and significant computational resources for model updates. Unlike biological brains, AI systems struggle with generalization and often require extensive retraining to adapt to new or slightly altered data inputs. The human brain consists of approximately 86 billion neurons, each forming synapses with thousands of other neurons, creating a massively parallel processing system. This structural complexity enables the brain to perform diverse and sophisticated functions, including pattern recognition, language processing, and sensory integration, with remarkable efficiency. While AI architectures like artificial neural networks mimic the brain's structure to some extent, they lack the same level of complexity and adaptability, constrained by hardware limitations and architectural rigidity. The energy inefficiency of digital AI is further exacerbated by the need for additional cooling systems to manage the heat generated by intensive computations. Understanding these processes not only sheds light on the nature of life itself but also provides insights into the evolution of biological and artificial efficiency, highlighting significant potential to improve the energy efficiency of digital technologies by adopting principles from biological systems.

5 CONCLUSION

In this paper, we have thoroughly examined the energy efficiency and effectiveness of biological systems in comparison to modern digital artificial intelligence (AI) systems. Our analysis reveals that life, as a physical process, defies the natural tendency towards entropy by employing mechanisms that optimize energy consumption. This principle of negentropy—where systems reduce their entropy by importing order from their environment—profoundly influences the efficiency of biological systems, from the simplest organisms like slime molds to complex beings with central nervous systems.

Biological entities exemplify remarkable energy optimization through various strategies. Simple organisms, such as slime molds, demonstrate sophisticated behaviors that minimize energy expenditure while solving complex problems, like finding the shortest path through a maze. In more advanced organisms, such as those with central nervous systems, energy optimization is achieved through intricate metabolic pathways, efficient signal transmission, and adaptive behaviors. These strategies include structural optimizations, functional tuning, behavioral adaptations, and social cooperation. For instance, the human brain, with its extensive network of neurons, operates with remarkable energy efficiency, performing complex cognitive tasks while consuming a fraction of the energy required by modern AI systems.

In contrast, current AI systems, particularly those involving deep learning and neural networks, exhibit significant inefficiencies. Despite technological advances, these systems consume energy at a magnitude far exceeding that of biological processes. The comparison highlights that AI systems are approximately 6-8 orders of magnitude less energy-efficient than their biological counterparts. This inefficiency is evident in the vast computational power required for AI training and operation, the large amounts of data needed, and the substantial energy consumption associated with cooling systems and hardware.

The disparity between biological and AI systems becomes even clearer when examining the limitations of current AI technologies. Modern AI algorithms, particularly those based on artificial neural networks (ANNs), lack optimization strategies such as neural network pruning and feedback loops. Additionally, AI systems do not incorporate the self-organizing principles inherent in biological systems, resulting in inefficiencies in learning and adaptation. The need for extensive data processing and retraining further exacerbates the energy inefficiency of AI systems.

Understanding these processes not only sheds light on the fundamental nature of life and its energy optimization mechanisms but also provides valuable insights into the evolution of biological and artificial efficiency. The principles of negentropy and energy optimization observed in biological systems offer potential pathways for enhancing the efficiency of AI technologies. By integrating these biological strategies, there is significant potential to improve the design and functionality of AI systems, moving towards more sustainable and effective computational technologies.

ACKNOWLEDGEMENT

Prof. Vallverdú research is supported by an ICREA Academia grant. Dr. Max Talanov is supported by the project "Artificial and bio-inspired networked intelligence for constrained autonomous devices – fairground" funded under the Piano Nazionale Ripresa e Resilienza (PNRR) Missione 4, "Istruzione e Ricerca" - Componente 2, "Dalla Ricerca all'Impresa" - Linea di Investimento 1.3, financed by the European Union – NextGenerationEU. The project "Future Artificial Intelligence – FAIR" has the code PE0000013 CUP (Master): J53C22003010006, CUP: J43C24000230007.

REFERENCES

[1] Andrew Adamatzky. (2010). *Physarum machines: computers from slime mould*, volume 74. World Scientific.

[2] David Attwell and Simon B Laughlin. (2001). An energy budget for signaling in the grey matter of the brain. *Journal of Cerebral Blood Flow & Metabolism*, 21(10):1133–1145.

[3] Alex de Vries. (2023). The growing energy footprint of artificial intelligence. *Joule*, 7(10):2191–2194.

[4] Claire Detrain, Jean-Louis Deneubourg, and Jacques M Pasteels. (1999). Decision-making in foraging by social insects. In *Information processing in social insects*, pages 331–354. Springer.

[5] Hamza Djigal, Jia Xu, Linfeng Liu, and Yan Zhang. (2022). Machine and deep learning for resource allocation in multi-access edge computing: A survey. *IEEE Communications Surveys & Tutorials*, 24(4):2449–2494.

[6] Brian D Fath. (2007). Network mutualism: Positive community-level relations in ecosystems. *Ecological modelling*, 208(1):56–67.

[7] Julia J Harris and David Attwell. (2012). The energetics of cns white matter. *Journal of Neuroscience*, 32(1):356–371.

[8] Julia J Harris, Renaud Jolivet, and David Attwell. (2012). Synaptic energy use and supply. *Neuron*, 75(5):762–777.

[9] Peter W Hochachka and George N Somero. (2002). *Biochemical adaptation: mechanism and process in physiological evolution*. Oxford university press.

[10] Mark Horowitz. (2014). 1.1 computing's energy problem (and what we can do about it). In *2014 IEEE international solid-state circuits conference digest of technical papers (ISSCC)*, pages 10–14. IEEE.

[11] Alexander Johnson, Julian Lewis, and Bruce ALBERTS. (2002). Molecular biology of the cell.

[12] Stuart A Kauffman. (1993). *The origins of order: Self-organization and selection in evolution*. Oxford University Press, USA.

[13] Simon B Laughlin. (2001). Energy as a constraint on the coding and processing of sensory information. *Current opinion in neurobiology*, 11(4):475–480.

[14] Seth Lloyd. (2002). Computational capacity of the universe. *Physical Review Letters*, 88(23):237901.

[15] Daniel R MacNulty, Aimee Tallian, Daniel R Stahler, and Douglas W Smith. (2014). Influence of group size on the success of wolves hunting bison. *PloS one*, 9(11):e112884.

[16] Eve Marder and Jean-Marc Goaillard. (2006). Variability, compensation and homeostasis in neuron and network function. *Nature Reviews Neuroscience*, 7(7):563–574.

[17] Richard Mayne and Andrew Adamatzky. (2015). Slime mould foraging behaviour as optically coupled logical operations. *International Journal of General Systems*, 44(3):305–313.

[18] Toshiyuki Nakagaki, Hiroyasu Yamada, and Ágota Tóth. (2000). Maze-solving by an amoeboid organism. *Nature*, 407(6803):470–470.

[19] Ali M Neishabouri and Aldo A Faisal. (2011). The metabolic efficiency of myelinated vs unmyelinated axons. *BMC neuroscience*, 12(Suppl 1):P100.

[20] David Patterson, Joseph Gonzalez, Quoc Le, Chen Liang, Lluis-Miquel Munguia, Daniel Rothchild, David So, Maud Texier, and Jeff Dean. (2021). Carbon emissions and large neural network training. *arXiv preprint arXiv:2104.10350.*

[21] Jürgen Schmidhuber. (2015). Deep learning in neural networks: An overview. *Neural networks*, 61:85–117.

[22] Erwin Schrödinger. (1944). What is life? the physical aspect of the living cell.

[23] Magnus Selin, Mattias Tiger, Daniel Duberg, Fredrik Heintz, and Patric Jensfelt. (2019). Efficient autonomous exploration planning of large-scale 3-d environments. *IEEE Robotics and Automation Letters*, 4(2):1699–1706.

[24] Daniel Sperling and Deborah Gordon. (2009). *Two billion cars: driving toward sustainability*. Oxford University Press.

[25] Vivienne Sze, Yu-Hsin Chen, Tien-Ju Yang, and Joel S Emer. (2017). Efficient processing of deep neural networks: A tutorial and survey. *Proceedings of the IEEE*, 105(12):2295–2329.

[26] Jordi Vallverdú, Oscar Castro, Richard Mayne, Max Talanov, Michael Levin, Frantisek Baluška, Yukio Gunji, Audrey Dussutour, Hector Zenil, and Andrew Adamatzky. (2018). Slime mould: the fundamental mechanisms of biological cognition. *Biosystems*, 165:57–70.

Study of Spontaneous Electrical Low-frequency Oscillations and Morphology of Mycelium Grown in Thermal Polypetide Media

DAVIN BROWNER*, PANAGIOTIS MOUKOGIANNIS AND ANDREW ADAMATZKY

Unconventional Computing Lab, UWE, Bristol, UK

Received: September 1, 2024. Accepted: September 28, 2024.

Fungi have existed on earth for at least hundreds of millions of years suggesting that mats and networks of mycelium could have formed an early source of protocognitive activity. Amino acids are a fungal cell wall permeable source of nutrition for many present day fungi and this relationship may have been initiated in certain early earth conditions fuelling metabolic processes that could not occur otherwise. Particular kinds of peptides called thermal proteins or proteinoids could have played an important role in this geological period due to their assembly at high temperatures (150+°C) and ability to form electrical junctions with biological organisms via attachment of microspheres. In artificial cultures proteinoids could be used as a method to interface with intracellular and extracellular electrophysiology. To investigate these conditions and the resulting dynamics of electrical signaling, we studied the characteristics of an in-vitro culture of mycelium cultivated with proteinoids. Scanning electron micrographs show growth of the basidiomycete *Schizophyllum commune* with proteinoid microspheres attached to hyphae suggesting that this fungal species grows successfully in proteinoid solution. Further, we show evidence for the existence of integration of electrical signaling between the abiotic peptides and fungus in via analysis of the spontaneous low frequency extracellular oscillations.

Keywords: Mycelium, thermal polypeptides, low frequency extracellular bioelectrical oscillations

1 INTRODUCTION

Fungi are ancient organisms that have undergone evolution for at least hundreds of millions of years [8, 20]. Mycorrhizal fungi have established

* Corresponding author: E-mail: davin.browner@uwe.ac.uk

symbiotic associations with a diverse array of organisms, including bacteria, plants, and animals, and have been instrumental in changing Earth's ecosystems [8, 39]. Fungi is one of the oldest known examples of a basally cognitive organism with a presence on earth for over a billion years [27]. Prior to the evolution of plants mycelium and fungal fruiting bodies evolved and were potentially sustained by bacteria, slime moulds or sources of nitrogen including amino acides, amines and amides [27]. Most fungi can assimilate amino acids, amines and amides as nitrogen sources fuelling metabolic processes that cannot occur otherwise [44]. Furthermore, but with the exception of many yeasts, fungi are also proteolytic and can hydrolyse proteins (via extracellularly secreted proteases) to liberate utilizable amino acids for growth [44]. Proteinoids may have formed previously in the hot salty volcanic conditions of a prebiotic earth and are a potential candidate for a partial role in supporting the metabolic activity of fungi during this period [36]. Therefore, proteinoids comprising of amino acid polymers formed following heating to above their melting point (150-200°C) are a strong candidate for sustenance and electrical signaling integration with primordial mycelial mats and networks via this metabolic pathway. In particular, the spontaneous low frequency oscillations ($\leq$ 1 hz) which have been shown to be an important part of bioelectrical integration and signalling over longer timescales (hours to days) relevant to morphological development of basally cognitive organisms [18].

What is common between proteinoids and fungi? Proteinoid microspheres could have been generated at the very early stages of life development on Earth [15]. These features make proteinoids and fungi the most unique representatives of early life forms. Basic communication between proteinoids and mycelium may have formed via spikes in electrical potential in response to extracellular cues and signal transduction methods, external chemical gradients, translocation of nutrients and growth processes. Spikes of electrical potential are well-known characteristics of neurons and are attributed to their learning and decision-making processes [4, 6, 37]. However, it is not just neurons that can generate spikes of electrical potential. Other living organisms, such as Protozoa [7, 14, 34], Hyrdoroza [18], slime molds [22, 24], and plants [17, 42, 47], also exhibit this capability. The occurrence of action potential-like spiking activity in fungi was first documented in 1976 [41], further confirmed in 1995 [33], and techniques for recording the electrical activity in fruiting bodies and colonized substrates were identified in 2018 [2]. Proteinoids and fungi are examples of transfer functions with broad biological significance. They are both instances of biological electronics that might lead to various applications in organic electronics. The ability of a heated protein to transmit an electrical signal from one molecule to another is known as its transfer function in proteinoids. Here, the thermal protein acts as a molecular adapter, binding to several molecules and transferring an electrical signal along an established pathway.

Elucidation of the low frequency electrical spiking activity in basal cognition can be aided through methods to record the extracellular potentials over longer timescales [28, 43]. While this is traditionally associated with bioelectricity in neural cultures, there exists opportunity to study electrical spiking and associated effects in aneural or non-neural systems including fungi, proteinoids, slime moulds and others. Non-neural organisms under different metabolic states induced by proteinoids are therefore a useful way to probe the conditions of the emergence of basal cognition and exploration of protobiological modifications of extracellular signaling. In this paper, we explore the characteristics of these basally cognitive systems through in-vitro study of mycelium and proteinoid cultures. To investigate these conditions and the resulting dynamics of low frequency signalling, we studied the characteristics of the culture of mycelium and proteinoids. Scanning electron micrographs show growth of the basidiomycete *Schizophyllum commune* with proteinoid microspheres attached to hyphae confirming that this fungal culture grows well in proteinoid solution. Further, we provided evidence for the existence of integration of electrical signalling between the abiotic peptides and fungus in the extracellular space of the cultures surveyed. We proposed that our proteinoid-fungal matrices' complex composition produces periodic electrical spikes in the 0.05-0.3 mV range resembling short bursts of slow oscillatory-like activity in a system lacking conventional cellular structure. The results section provides an in-depth investigation of these electrical potential spikes, verifying their existence and uncovering complex patterns of activity throughout the interconnected network. In addition, the results section explores the fractal dimensions, entropy, and dimensionality metrics of the spiking activity, uncovering different levels of complexity and synchronisation among channels. Using methods from information theory, like mutual information analysis, can help us understand how proteinoid and fungus communicate and what role certain channels play in processing information.

METHODS

Proteinoids preparation

L-Aspartic acid and L-Glutamic acid amino acids from Sigma Aldrich were used to synthesize proteinoids via thermal polycondensation invented in the 1960s [16]. The amino acids were heated at 180°C for 30 minutes in a nitrogen atmosphere with continuous stirring. The resulting proteinoids were lyophilized and stored at room temperature.

Preparation of mycelium

The H4-8A strain of the split-gill fungus Schizophyllum commune was grown on Schizophyllum commune minimal medium (SCMM) with 1.5% agar for three days at 30°C in the dark.

Preparation of mycelium-proteinoid matrices

Mycelium-proteinoid matrix preparation: After 7-8 days of growth and electrode insertion, 2ml of proteinoid solution was injected into the fungal culture. This allows damaged hyphae to regrow and attach to the electrode surface. SEM image of this process and its structural features is shown in Figure 1.

Scanning electron microscopy (SEM)

An FEI Quanta 650 microscope was used to study the morphology of proteinoids. Samples were gold sputter-coated to increase conductivity for SEM imaging. Proteinoids are intrinsically conductive, even in high-vacuum environments, and can be imaged without conductive coating. However, fungal samples required gold sputtering due to reduced conductivity caused by dehydration during preparation.

Electrical recordings of spontaneous low frequency oscillations

Proteinoids are difficult to study using conventional intracellular probing methods such as patch clamp. This common method is ineffectual and cannot form a seal on the surface of the microspheres. Moreover, there is no known method for successfully piercing the membrane of the proteinoid without rendering the microspheres electrically inactive or indistinguishable from the volume media. Therefore, macroscale electrical activity of the fungal colonies was recorded using pairs of sub-dermal needle electrodes (Spes Medica S.r.l., Italy) coated with iridium on stainless steel. The electrodes were connected to a Pico Technology ADC-24 high-resolution data logger with a 24-bit analog-to-digital converter, ensuring noise-free resolution through galvanic isolation and adjustable sample rates. The distance between electrodes in each differential pair was approximately 10 mm. Electrodes were placed radially outward from the inoculation point, with 8 individual electrodes or 4 differential pairs (Ch1-4). To maintain colony integrity, Petri dishes were not opened during experiments, and recordings were conducted after 7-8 days of growth to allow mycelium regrowth. Small openings were made in the Petri dish lids using a hot needle, and electrodes were inserted through these openings until they reached the cultures. The tips of the electrodes were isolated via tubing fitted to main shaft. All recordings were performed simultaneously and lasted over a month.

RESULTS

The scanning electron micrographs depicted in Figure 1 offer visible confirmation of the close contact between proteinoids and fungus at the nanoscale. These photos provide evidence that proteinoids and fungus have co-evolved and formed complex interactions that could have had a role in the development of intelligence. The micrographs demonstrate the existence of proteinoid nanoparticles that are linked to the surface of the fungal hyphae. Significantly, the proteinoid nanoparticles exhibit a gradual rise in size, with the largest measured nanosphere being 500 nm in diameter. The progressive enlargement indicates a growth mechanism or the accumulation of proteinoid nanoparticles on the fungal surface. Moreover, the detection of proteinoid nanoparticles adhering to various sections of the same hyphae suggests a widespread and maybe specific interaction between the proteinoids and the fungal mycelium. This close link suggests a potential functional relationship or symbiotic interaction between the two organisms. The micrographs were captured using a high magnification of 30,521x and a low voltage of 2.00 kV, which enables a detailed visualisation of the proteinoid-mycelium interface. The 348 nm scale bar serves as a visual guide to indicate the size of the observed structures, highlighting the fact that this interaction occurs at the nanoscale. The gradual increase in proteinoid nanoparticles on the fungal surface suggests a dynamic process of interaction and potential exchange of information or materials between the two species.

The voltage oscillations observed across the mycelium–proteinoid hybrid structure show complex and fluctuating patterns over space and time, as illustrated in Figure 2. By recording the signals from four separate channels, it is possible to determine the origin of specific rhythms or spikes, whether they are generated locally or propagate from distant regions within the interconnected conductive network. Statistical analysis of each channel reveals further variability in the signal profiles. Channel 1 exhibits an average potential close to zero, but the recorded values span a range of over 1 mV at the extremes. In comparison, Channel 2 has a considerably higher average potential, while Channel 3 displays more negative values at its lowest points. To make sense of these complex signal streams, it is necessary to first disentangle the intricate mixing of multiple signaling inputs across the heterogeneous and organised landscape. Mapping these emergent bio–synthetic matrices is an initial step towards integrating hybrid living electronics and extracting meaningful patterns from the inherent noise. Statistical analysis of the peak spike amplitudes detected in each channel provides information about the signal characteristics (Table 1). The highest average spike amplitude of 0.367 mV was observed in Channel 3, while Channels 1, 2, and 4 showed average amplitudes ranging from 0.1 to 0.15 mV. This varia-

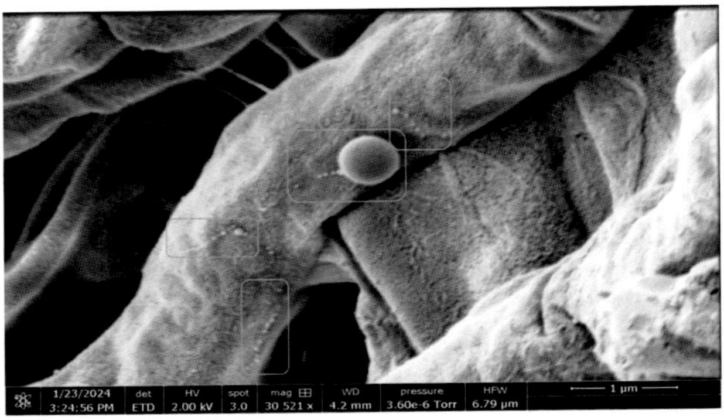

FIGURE 1
Scanning electron micrograph of proteinoid-mycelium interface showing: (a) an aggregation of proteinoids attached to the surface of the hyphae with increasing size culminating in a proteinoid with 500 nm diameter. Proteinoid nanoparticles can be seen attached to multiple regions of the same hyphae. Operating parameters were: magnification of 30,521x, voltage of 2.00 kV. and scale bar of 1 micrometer (μm); (b) three nanospheres in close proximity attached to the surface of the mycelium. Operating parameters were: magnification of 33,765x, voltage of 2.00 kV and scale bar of 1 μm. Locations of the microspheres are indicated by green outlines.

tion in spike amplitudes across different channels suggests that spatial factors related to the structure of the hybrid material influence the signal properties. The composite material displays frequency–dependent behaviour, with specific channels exhibiting distinct electrical properties. The attenuation spectra reveal the complex transmission dynamics within the material, which are influenced by its diverse internal structure. The attenuation levels vary from -68 to over -100 dB at frequencies ranging from 0 to 5 Hz, depending on the channel observed. Channel 1 exhibits high–pass filtering characteristics, where the attenuation decreases with increasing frequency, indicating a preference for the conduction of faster impulses. In contrast, Channel 2 shows a unique preference for low to medium frequencies, characterized by step–like decreases in attenuation that do not follow a monotonic trend. The presence of multiple frequency signaling motifs within the dense bio–electric network suggests a high computational capacity for encoding spectral information. This is achieved by routing inputs through specific pathways that are optimized for desired band–limited temporal coding schemes. Ongoing research aims to further elucidate the intricate relationships between the structure and function that govern the formation of specific conduction bands. The precise organisation of the interconnected microbiological, proteinoid, and inorganic components has the potential to create specialized bio–electric fabrics capable of efficiently transmitting precisely timed codes. The electrical potential spikes recorded across the channels of the proteinoid–fungal matrix exhibit

Channel	Mean (mV)	Standard Deviation (mV)	Minimum (mV)	Maximum (mV)
Ch1	0.145149	0.035086	0.100300	0.376540
Ch2	0.116457	0.020281	0.100130	0.215660
Ch3	0.367876	0.131160	0.102200	2.064940
Ch4	0.122516	0.030273	0.100050	0.344980

TABLE 1
This table shows the peak statistics for electrical potentials of each microelectrode channel (Ch1, Ch2, Ch3, and Ch4) measured in millivolts (mV). The listed statistical parameters include the mean, which gives an overall average level of the signal's peaks; standard deviation, showing the variation or dispersion from the average; and the minimum and maximum recorded peaks in each channel, providing an idea of the range of potential values detected. These peak statistical data offer vital insights into the properties and behaviour of the recorded electrical signals.

centralized tendencies that can be fitted to Gaussian distributions. The transient bursts follow normal distributions, which can be described by the probability density function: $f(x|\mu, \sigma^2) = \frac{1}{\sigma\sqrt{2\pi}} e^{-\frac{(x-\mu)^2}{2\sigma^2}}$ where μ represents the mean spike amplitude and σ denotes the standard deviation for each channel. Channel 3 exhibited the highest average spike magnitude at 0.367 mV, along with a larger variability characterized by σ = 0.131 mV. In contrast, Channel 2 showed the most tightly regulated spiking behavior, with a mean amplitude of μ = 0.116 mV. These differences in the statistically modeled spiking behavior are likely attributed to the heterogeneity in structural factors and molecular transport across the internal domains within the co–assembled biomolecular matrix.

The electrical potential spikes detected across the proteinoid–fungal matrix channels exhibit centralized tendencies amenable to Gaussian distribution fitting. The transient bursts empirically follow normal distributions, and the highest average spike magnitude of 0.367 mV emerged in Channel 3 with a broader variability of σ = 0.131 mV. In contrast, Channel 2 demonstrated the most tightly regulated spiking around μ = 0.116 mV.

Fractal Dimension Analysis of Mycelium–Proteinoid Spiking Data in Four Channels

We analyzed the fractal dimensions of the resulting signals using multiple approaches: Higuchi (D_H) [21], Katz (D_K) [25], and Petrosian (D_P) [35].

Higuchi Method (D_H)

The Higuchi method estimates the length of the graph as a function of the indices difference (equivalent to differences in t). Given a time series $X : 1, \ldots, N \to \mathbb{R}$ consisting of N data points and a parameter $k_{max} \geq 2$, the Higuchi Fractal Dimension (HFD) of X is calculated as the average length

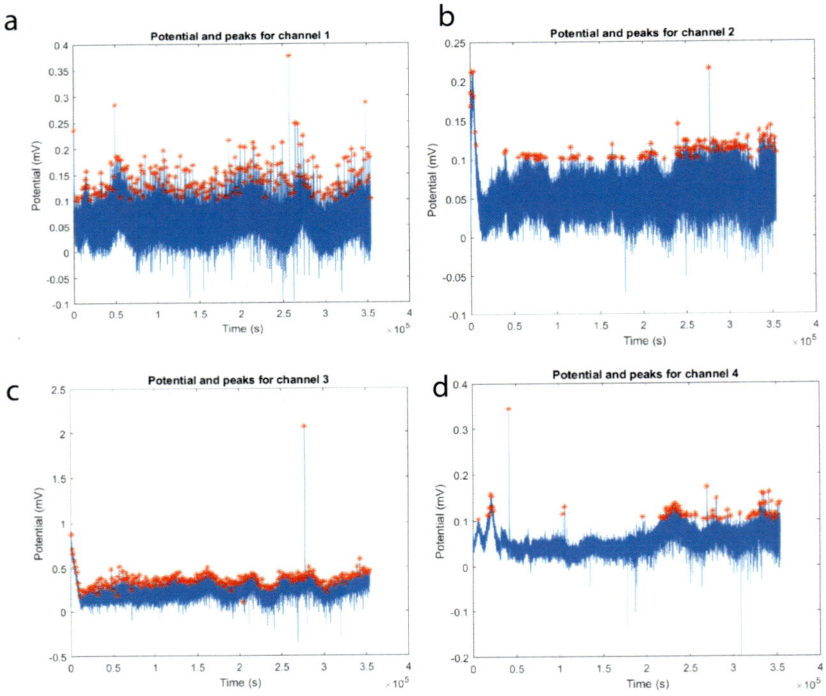

FIGURE 2
a) Distribution and statistics of identified maximum electrical potentials for Channel 1 after eliminating background interference. The average maximum value is 0.145 mV, indicating a moderate amount of activity in this channel. b) This figure presents a summary of the peak potentials detected in Channel 2 after the removal of background noise. The mean maximum amplitude recorded is 0.116 mV, which is comparatively lower than that of other channels. c) Depiction of peaks and corresponding statistical data for Channel 3 following the elimination of background noise. This channel demonstrates increased activity, with a mean peak value of 0.367 mV. d) This diagram illustrates the dispersion of maximum potentials for Channel 4, after the elimination of background noise. The average maximum amplitude is 0.122 mV, which is similar to the amplitudes observed in Channels 1 and 2.

versus k:

$$L_m(k) = \frac{N-1}{\left\lfloor \frac{N-m}{k} \right\rfloor k^2} \sum_{i=1}^{\left\lfloor \frac{N-m}{k} \right\rfloor} |X_N(m+ik) - X_N(m+(i-1)k)| \tag{1}$$

$$L(k) = \frac{1}{k} \sum_{m=1}^{k} L_m(k) \tag{2}$$

Using k_{max}‘ $= 10$, Ch1 ($D_H = 1.8$), Ch2 ($D_H = 1.8$), and Ch4 ($D_H = 1.7$) show higher summed D_H values compared to Ch3 ($D_H = 1.4$). Ch3 recorded a drop in piecewise values between piece indices 10 and 50, indicating increasing self-similarity and decreasing complexity. Ch1 piecewise values followed a trend of decreasing self-similarity with increasing complexity values. Ch2 and Ch4 both showed piecewise progression of decreasing complexity and increasing self-similarity.

Katz Method (D_K)

Katz's method calculates the fractal dimension of a sample as follows: the sum and average of the Euclidean distances between the successive points of the sample (L and a, respectively) are calculated, as well as the maximum distance between the first point and any other point of the sample (d). The fractal dimension of the sample (D) is then:

$$D = \frac{\log_{10}(L/a)}{\log_{10}(d/a)} = \frac{\log_{10}(n)}{\log_{10}(d/L) + \log_{10}(n)} \tag{3}$$

where n is L divided by a.

Using an axis variable of -1, Ch1 ($D_K = 4.3$), Ch2 ($D_K = 3.9$), and Ch4 ($D_K = 3.4$) show higher summed D_K values compared to Ch3 ($D_K = 3.0$). The progression of piecewise values for D_K on all channels followed a similar pattern to D_H, with increasing self-similarity and lowering complexity over the time series for Ch1, decreasing self-similarity and increasing complexity for Ch2 and Ch4. Ch3 showed increasing self-similarity and decreasing complexity but with a lower mean variance compared to the same channel D_H.

Petrosian Method (D_P)

The Petrosian fractal dimension of a time series x is defined by:

$$P = \frac{\log_{10}(N)}{\log_{10}(N) + \log_{10}\left(\frac{N}{N+0.4N_\delta}\right)} \tag{4}$$

where N is the length of the time series, and N_δ is the number of sign changes in the signal derivative.

To further characterize the spiking activity, entropy and dimensionality metrics were calculated for each channel, as summarized in Table 2.

Channel 1 exhibited the highest spectral entropy (0.896) and Higuchi fractal dimension (D_H = 1.885), indicating a higher degree of complexity and irregularity in its spiking patterns. In contrast, Channel 3 showed the lowest spectral entropy (0.667) and Higuchi fractal dimension (D_H = 1.492), suggesting more regular and predictable spiking behaviour.

Channel	Spectral Entropy	D_H	D_K	D_P
Ch1	0.896	1.885	4.344	1.01043
Ch2	0.813	1.869	3.950	1.01039
Ch3	0.667	1.492	3.019	1.00957
Ch4	0.803	1.715	3.404	1.01044

TABLE 2
Entropy and Dimensionality Metrics of Spike Activity

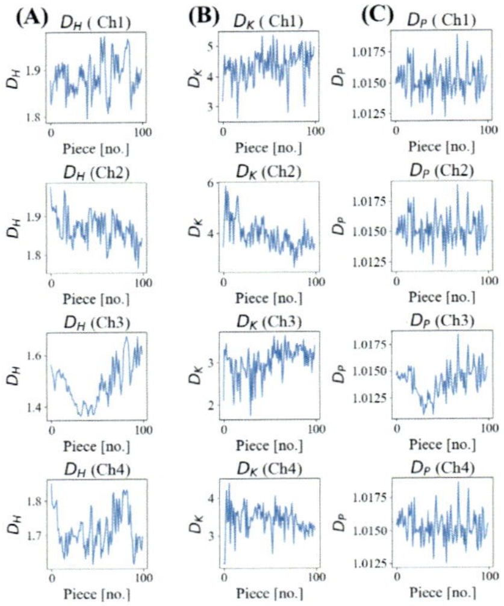

FIGURE 3
Fractal dimensions of each channel showing (a) Higiuchi dimensions (D_H) (b) Katz dimensions (D_K) and (c) Petrosian dimensions (D_P) plotted with the piecewise indices.

Channel Pair	Peak Correlation	Lag
Ch1 – Ch2	1224.3417	−2
Ch1 – Ch3	5188.3629	0
Ch1 – Ch4	1214.2090	−1
Ch2 – Ch3	5231.5419	+2
Ch2 – Ch4	1164.3559	0
Ch3 – Ch4	4898.9457	−1

TABLE 3
Peak synchronized activity between channel spike time series.

Cross–correlation analysis was conducted to investigate the synchronization and temporal relationships between the spiking activity of different channels. The peak correlation values and corresponding lags are summarized in Table 3. Channel pairs 1–3 and 2–3 exhibit the highest peak correlations (5188.36 and 5231.54, respectively) at lags of 0 and +2, indicating strong synchronization between these channels.

These results demonstrate that the proteinoid–fungal matrix exhibits complex and heterogeneous spiking activity, with varying degrees of complexity, synchronization, and information flow between channels. The observed patterns and relationships provide insights into the underlying mechanisms and interactions within the system, paving the way for further investigations into the functional significance of these spiking dynamics.

Mutual Information Analysis of Mycelium–Proteinoid Spiking Data in Four Channels

We gained valuable insights into the complex dynamics of proteinoid-fungus communication by applying an information theory approach [11, 26, 38, 40, 46]. This method unveiled the underlying factors driving the observed spiking behaviors. We computed self-information and mutual information across four channels. Based on the correlations found in their respective spiking activities, we estimated each channel's capacity to encode information and the degree of interaction between them. The resulting bit values quantify the reduction in uncertainty about the signals in one channel that can be obtained by observing another.

Mutual information represents the amount of information about one channel that can be learned from observing another. Our results show that the diagonal values are the highest, indicating that observing a channel yields the most information about itself. The self-information for channels 1, 2, 3, and 4 is as follows: $I(C_1;C_1) = 13.2044$, $I(C_2;C_2) = 13.0106$, $I(C_3;C_3) = 14.6916$, and $I(C_4;C_4) = 13.0178$. Notably, channel 3 had the highest self-information, indicating a larger information content from this channel. We also found varying degrees of mutual information between different channels, indicating their level of interaction. For example, $I(C_1;C_2) = 7.7863$ and $I(C_1;C_3) = 9.4629$ show that there is more mutual interaction between proteinoids and fungi in channels 1 and 3 than in channels 1 and 2. Focusing on channel 3 with others reveals increased mutual interaction, with values $I(C_4;C_3) = 9.2770$, $I(C_2;C_3) = 9.2697$, and $I(C_1;C_3) = 9.4629$. These findings are shown in Table 4. These results suggest that channel 3 plays a crucial role in the spiking activity of proteinoids and fungi. This implies that Channel 3 acts as a major hub of interaction within the dynamic assembly matrix. Our study provides a comprehensive understanding of the kinetics of information processing between proteinoids and fungus, emphasizing each channel's independent contributions and their interrelated roles. The findings could have far-reaching implications for modulating the environment to optimize this interaction.

	Ch1	Ch2	Ch3	Ch4
Ch1	13.2044	7.7863	9.4629	7.7933
Ch2	7.7863	13.0106	9.2697	7.6010
Ch3	9.4629	9.2697	14.6916	9.2770
Ch4	7.7933	7.6010	9.2770	13.0178

TABLE 4
Mutual Information Between Channels

Cross–Correlation Analysis of Channel Interactions in Fungi–Proteinoid Spiking Activity

Cross-correlation analysis provides an effective approach for quantifying similarities in the spike timing between channels as a function of lag:

$$R_{xy}(\tau) = \int_{-\infty}^{\infty} x(t)y(t+\tau)dt \tag{5}$$

Where $x(t)$ and $y(t)$ refer to the time-dependent signals from each channel, and τ is the offset lag time.

Multiple channel pairings exhibit strong cross-correlation exceeding 1200, with Channel 3 highly correlated to all zones at various lags. The peak correlation of 5231 between Channels 2 and 3 at a +2 lag implies tight temporal coupling, with activity in Channel 3 preceding Channel 2. Overall, the significant cross-talk confirms interdependent spiking relationships across the ensemble.

Quantitative cross-correlation analysis confirms the presence of substantial interrelationships between localized spiking behaviors within the integrated proteinoid-fungal matrix (Table 3). Multiple channel pairs have correlation peaks that exceed 1000, with Channels 2 and 3 having the highest level at nearly 5200. The greatest correlation delays exhibit both positive and negative offsets, suggesting complex dynamics with activity in specific zones preceding or following activity in other zones.

DISCUSSION

This study investigates the connection between mycelial fungi and proteinoids as representations of early cognitive processes [5, 19, 29, 31, 32] that do not rely on conventional neural structures [9]. Extracellular electrical potential spiking that resembles bursts of neural–like spiking was uncovered by connecting these two platforms of intricate self–organized biomolecular dynamics [13]. Proteinoid–mycelia composites exhibit sensory input

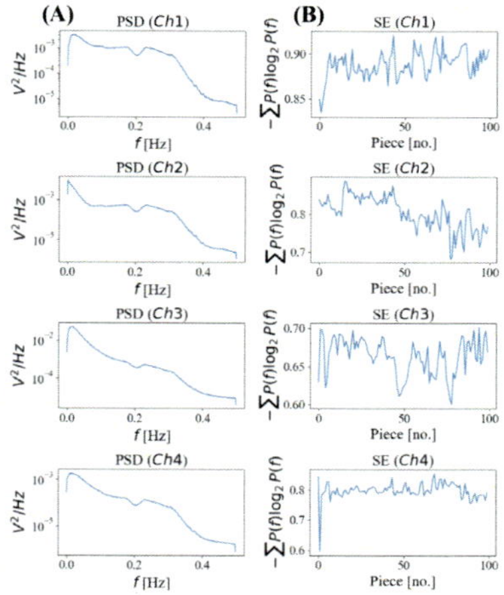

FIGURE 4
(a) The per channel power spectral density (V^2/Hz) plotted with frequency f [Hz] (b) spectral entropy ($-\sum P(f)\log_2 P(f)$) for each channel plotted with the piecewise indices.

coupling, nonlinear stimulus filtering similar to logic gating, and low frequency electrical dynamics that result in lateral inhibition effects. These effects resemble primitive processing and decision thresholding functionalities [3, 30]. The experimental system used to observe these phenomena is simple and can be improved through use of microelectrodes or dense arrays. However, using a simple system, the results uncover a convincing biochemical foundation for the emergence of rhythmic signalling as a result of mycelium growth in solutions of proteinoid [12] and integration of sensory information that could have enhanced the combined intelligence of communal living chemical structures on ancient Earth.

In addition to its implications for the origins of life, the study provides solid evidence of hybrid bio–abiotic materials [45] that display complex and spontaneous signal changes. Proteinoid structures on the surface of hyphae may present opportunities for bioelectrical interfacing with the intracellular environment of individual hyphae. In addition, proteiniods resemble attachment methods observed in bacteria on the surface of hyphae and may be useful as analogs in such studies. This study provides a basic exploration of the wider range of possibilities in combining synthetic polypeptides, organic electrical scaffolds, and living matrices to create a diverse array of adjustable intelligent substrates [10]. In addition to their effects on the origins of life, the observed oscillatory behaviours, which are influenced by sensory input, provide more evidence of the potential for fungal–bacterial assemblies [23] in unconventional computer applications [1].

CONCLUSIONS

To summarise, the complex structure of our proteinoid–fungal matrix generates extracellular low frequency electrical spikes in the range of 0.05-0.3 mV. These spontaneous oscillations resemble long term biolelectrical states in other organisms (e.g., hydra [18]). An objective study of the magnitudes highlights the widespread nature of signalling activity in this dynamic network, where specific regions achieve maximum potentials of around 0.37 mV. The presence of a distinct biophysical basis in our work is highlighted by the pulsing behaviour, which may enable communication and the creation of information patterns inside these communal mats.

FUNDING

PM and AA were supported by EPSRC Grant EP/W010887/1 "Computing with proteinoids. DB is funded under the framework of the FUNGATERIA (www.fungateria.eu) project, which has received funding from the European Unions HORIZON-EIC-2021-PATHFINDER CHALLENGES programme under grant agreement No. 101071145. It is co-funded by the UK Research and Innovation grant No. 10048406.

CONFLICTS OF INTEREST

The authors declare no competing interests.

ACKNOWLEDGEMENTS

The authors are grateful to David Patton for helping with SEM imaging.

DATA AVAILABILITY

REFERENCES

[1] Andrew Adamatzky, (2014). Unconventional computing.

[2] Andrew Adamatzky. (2018). On spiking behaviour of oyster fungi pleurotus djamor. *Scientific reports*, 8(1):1–7.

[3] Andrew Adamatzky. (2021). Towards proteinoid computers. hypothesis paper. *Biosystems*, 208:104480.

[4] Daniela Sabrina Andres. (2015). The language of neurons: theory and applications of a quantitative analysis of the neural code. *International Journal of Medical and Biological Frontiers*, 21(2):133.

[5] William Bains and Dirk Schulze-Makuch. (2016). The cosmic zoo: the (near) inevitability of the evolution of complex, macroscopic life. *Life*, 6(3):25.

[6] Morris H Baslow. (2009). The languages of neurons: an analysis of coding mechanisms by which neurons communicate, learn and store information. *Entropy*, 11(4):782–797.

[7] MS Bingley. (1966). Membrane potentials in amoeba proteus. *Journal of Experimental Biology*, 45(2):251–267.

[8] Meredith Blackwell. (2011). The fungi: 1, 2, 3 5.1 million species? *American journal of botany*, 98(3):426–438.

[9] David Bodian. (1962). The generalized vertebrate neuron: Conventional terminology of neuron structure lags behind current functional-anatomical concepts. *Science*, 137(3527):323–326.

[10] Marinos Charalambides, George Pavlou, Paris Flegkas, Ning Wang, and Daphne Tuncer. (2011). Managing the future internet through intelligent in-network substrates. *IEEE Network*, 25(6):34–40.

[11] Thomas M Cover. (1999). *Elements of information theory*. John Wiley & Sons.

[12] Koen de Reus, Masayo Soma, Marianna Anichini, Marco Gamba, Marianne de Heer Kloots, Miriam Lense, Julia Hyland Bruno, Laurel Trainor, and Andrea Ravignani. (2021). Rhythm in dyadic interactions. *Philosophical Transactions of the Royal Society B*, 376(1835):20200337.

[13] Rashmi C Desai and Raymond Kapral. (2009). *Dynamics of Self-organized and Self-assembled Structures*. Cambridge University Press.

[14] R Eckert, Y Naitoh, and K Friedman. (1972). Sensory mechanisms in paramecium. i. *J. exp. Biol*, 56:683–694.

[15] Sidney W Fox. (1992). Thermal proteins in the first life and in the "mind-body problem. In *Evolution of Information Processing Systems*, 203–228. Springer.

[16] Sidney W Fox and Kaoru Harada. (1960). The thermal copolymerization of amino acids common to protein1. *Journal of the American Chemical Society*, 82(14):3745–3751.

[17] Jörg Fromm and Silke Lautner. (2007). Electrical signals and their physiological significance in plants. *Plant, cell & environment*, 30(3):249–257.

[18] Alison Hanson. (2021). Spontaneous electrical low-frequency oscillations: a possible role in hydra and all living systems. *Philosophical Transactions of the Royal Society B*, 376(1820):20190763.

[19] H Hartman. (1982). Life, language, and society.

[20] David L Hawksworth. (2001). The magnitude of fungal diversity: the 1· 5 million species estimate revisited. *Mycological research*, 105(12):1422–1432.

[21] Tomoyuki Higuchi. (1988). Approach to an irregular time series on the basis of the fractal theory. *Physica D: Nonlinear Phenomena*, 31(2):277–283.

[22] Tatsuichi Iwamura. (1949). Correlations between protoplasmic streaming and bioelectric potential of a slime mold, *Physarum polycephalum*. *Shokubutsugaku Zasshi*, 62(735-736):126–131.

[23] Shuo Jiao, Ziheng Peng, Jiejun Qi, Jiamin Gao, and Gehong Wei. (2021). Linking bacterial-fungal relationships to microbial diversity and soil nutrient cycling. *Msystems*, 6(2):e01052–20.

[24] Noburo Kamiya and Shigemi Abe. (1950). Bioelectric phenomena in the myxomycete plasmodium and their relation to protoplasmic flow. *Journal of Colloid Science*, 5(2):149–163.

[25] Michael J Katz. (1988). Fractals and the analysis of waveforms. *Computers in biology and medicine*, 18(3):145–156.

[26] Joseph T Lizier, Jakob Heinzle, Annette Horstmann, John-Dylan Haynes, and Mikhail Prokopenko. (2011). Multivariate information-theoretic measures reveal directed information structure and task relevant changes in fmri connectivity. *Journal of computational neuroscience*, 30:85–107.

[27] Corentin C Loron, Camille François, Robert H Rainbird, Elizabeth C Turner, Stephan Borensztajn, and Emmanuelle J Javaux. (2019). Early fungi from the proterozoic era in arctic canada. *Nature*, 570(7760):232–235.

[28] R Mayne and A Adamatzky. (2015). Slime mould foraging behaviour as optically coupled logical operations. *International Journal of General Systems*, 44(3):305–313.

[29] Leonid L Moroz, Mikhail A Nikitin, Pavlin G Poličar, Andrea B Kohn, and Daria Y Romanova. (2021). Evolution of glutamatergic signaling and synapses. *Neuropharmacology*, 199:108740.

[30] Tadayoshi Nakashima. (2005). Metabolism of proteinoid microspheres. *Organic Geo-and Cosmochemistry*, 57–81.

[31] Anna Nikolaidou, Panagiotis Mougkogiannis, and Andrew Adamatzky. (2023). Electroactive composite biofilms integrating kombucha, chlorella and synthetic proteinoid proto–brains.

[32] Anna Nikolaidou, Panagiotis Mougkogiannis, and Andrew Adamatzky. (2023). Living kombucha electronics with proteinoids.

[33] S Olsson and BS Hansson. (1995). Action potential-like activity found in fungal mycelia is sensitive to stimulation. *Naturwissenschaften*, 82(1):30–31.

[34] Shunsuke Ooyama and Tatsuo Shibata. (2011). Hierarchical organization of noise generates spontaneous signal in paramecium cell. *Journal of theoretical biology*, 283(1):1–9.

[35] Arthur Petrosian. (1995). Kolmogorov complexity of finite sequences and recognition of different preictal eeg patterns. In *Proceedings eighth IEEE symposium on computer-based medical systems*, 212–217. IEEE.

[36] Kristof Plankensteiner, Hannes Reiner, and Bernd M Rode. (2006). Amino acids on the rampant primordial earth: electric discharges and the hot salty ocean. *Molecular Diversity*, 10:3–7.

[37] J Andrew Pruszynski and Joel Zylberberg. (2019). The language of the brain: real-world neural population codes. *Current opinion in neurobiology*, 58:30–36.

[38] Rodrigo Quian Quiroga and Stefano Panzeri. (2009). Extracting information from neuronal populations: information theory and decoding approaches. *Nature Reviews Neuroscience*, 10(3):173–185.

[39] Dirk Redecker, Arthur Schüßler, Herbert Stockinger, Sidney L Stürmer, Joseph B Morton, and Christopher Walker. (2013). An evidence-based consensus for the classification of arbuscular mycorrhizal fungi (glomeromycota). *Mycorrhiza*, 23:515–531.

[40] Thomas Schreiber. (2000). Measuring information transfer. *Physical review letters*, 85(2):461.

[41] Clifford L Slayman, W Scott Long, and Dietrich Gradmann. (1976). "Action potentials" in *Neurospora crassa*, a mycelial fungus. *Biochimica et Biophysica Acta (BBA) — Biomembranes*, 426(4):732–744.

[42] Kazimierz Trebacz, Halina Dziubinska, and Elzbieta Krol. (2006). Electrical signals in long-distance communication in plants. In *Communication in plants*, 277–290. Springer.

[43] Jordi Vallverdú, Oscar Castro, Richard Mayne, Max Talanov, Michael Levin, Frantisek Baluška, Yukio Gunji, Audrey Dussutour, Hector Zenil, and Andrew Adamatzky. (2018). Slime mould: the fundamental mechanisms of biological cognition. *Biosystems*, 165:57–70.

[44] Graeme M Walker and Nia A White. (2017). Introduction to fungal physiology. *Fungi: biology and applications*, 1–35.

[45] Xiaoliang Wang, Jianxu Zhang, Youjun Wang, Yang Liu, Zhenhua Li, Zhigang Xie, and Zhijun Chen. (2017). Fluorescent protein nanovessels: A new platform to generate bio–abiotic hybrid materials for bioimaging. *Advanced Functional Materials*, 27(43):1702051.

[46] Michael Wibral, Nicolae Pampu, Viola Priesemann, Felix Siebenhühner, Hannes Seiwert, Michael Lindner, Joseph T Lizier, and Raul Vicente. (2013). Measuring information-transfer delays. *PloS one*, 8(2):e55809.

[47] Matthias R Zimmermann and Axel Mithöfer. (2013). Electrical long-distance signaling in plants. In *Long-Distance Systemic Signaling and Communication in Plants*, 291–308. Springer.

Kombucha Mats as Responsive Materials

Anna Nikolaidou[1,4,*], Panagiotis Mougkogiannis[1], Alessandro Chiolerio[2,1], Mohammad Mahdi Dehshibi[3,1] and Andrew Adamatzky[1]

[1]*Unconventional Computing Laboratory, University of the West of England, Bristol, BS16 1QY, UK*
[2]*Bioinspired Soft Robotics, Istituto Italiano di Tecnologia, Via Morego 30, 16163 Genova, Italy*
[3]*Escuela Politécnica Superior, Departamento de Informática, Universidad Carlos III de Madrid, Leganés, 28911, Spain*
[4]*School of Architecture and Environment, University of the West of England, Bristol, BS16 1QY, UK*

Received: September 1, 2024. Accepted: September 28, 2024.

Responsive materials have attracted considerable interest because of their capacity to modify and alter their properties in reaction to external stimuli. This research explores the possibility of kombucha mats, which are a byproduct of the kombucha fermentation process, as a new and responsive material. By regulating the fermentation conditions and post-processing methods, one can adjust the distinctive electrical, mechanical and optical properties of kombucha mats. In this study, we employ mechanical testing, electrical measurements and optical stimulation to evaluate the responsive behaviour of kombucha mats. We examine the potential applications of kombucha mats as responsive materials, focusing on smart textiles, wearable devices, biosensors, soft robotics, and biomedical devices. We provide multiple case studies that demonstrate the successful incorporation of kombucha mats into practical and adaptable systems.This study demonstrates that kombucha mats have great potential as a sustainable and effective platform for creating responsive materials with specific qualities and a wide range of uses.

Keywords: Yeasts, bacteria, kombucha, electrical activity

1 INTRODUCTION

The development of stimuli-responsive materials from living biological molecules is an emergent area of research because of their great potential

* Corresponding author: E-mail: anna.nikolaidou@uwe.ac.uk

in diverse applications. In this study, we employ living kombucha cellulose mats as scaffold matrices to develop electrically responsive materials to mechanical, optical and electrical stimulation. We combine the properties of kombucha pellicles with the functionalities of targeted nanomaterials to create electroactive materials that can sense, process information and perform computation. Kombucha is a multispecies system derived from a symbiotic culture of yeast and bacteria. The term symbiosis is used to describe the metabolic interplay between the yeast and bacteria [59] and their close association in growth [19].

The origins of kombucha are believed to be traced back to Northeast China around 220 BC, with its consumption being prevalent during the Qin Dynasty [16]. The term "Kombucha" may have derived from the Japanese words for seaweed, "kombu," and tea, "Chao" [30]. The dominant microorganisms in kombucha are acetic acid bacteria [31], with *Acetobacter xylinum* being the most commonly referenced bacterium in the kombucha consortium, and *Schizosaccharomyces pombe* being the most frequently mentioned yeast [30].

Kombucha is produced through the fermentation of sugared tea using a symbiotic culture of bacteria and yeast known as SCOBY [15]. During the fermentation process, sucrose is converted into glucose and fructose [54]. The prominent microorganisms present in kombucha are acetic acid bacteria, which initially convert ethanol to acetaldehyde and then to acetic acid. These bacteria also catalyse the oxidation of glucose to gluconic acid [17,31]. The presence of acetic acid stimulates the production of ethanol, which, in turn, supports the growth of acetic acid bacteria and the production of acetic acid [35]. The microorganisms utilize sucrose as their primary carbon source, while the tea extract serves as the nitrogen source. In the presence of oxygen, SCOBY generates organic acids, carbon dioxide, and a cellulose-based mat [57]. Oxygen plays a crucial role in the production of kombucha mats, affecting the dynamics and behaviours of the yeast and bacteria [58,59]. Caffeine and related compounds such as theophylline and theobromine present in the tea infusion promote cellulose production by activating cellulosic complexes in the bacteria [6]. Additionally, bacterial cells are stimulated by vitamins and other nutrients released due to the death and autolysis of the yeast cells [8].

The fermentation process of kombucha typically lasts for about 14 days. At this stage, most of the carbon sources are used up, causing the fermentation to slow down [23, 24, 37]. Refreshing the sugared tea infusion can help prolong the fermentation process. After fermentation, the tea is composed of a cellulose mat floating on the surface and a liquid broth underneath [10]. The dominant microbes in the kombucha mat generally remain consistent throughout fermentation, although some changes in the yeast community have been observed, with no reported changes in the bacterial community [8,56]. Chemical analysis of kombucha cellulose mats has shown the presence of ethanol, various acids, carbohydrates, vitamins, and minerals [19,31].

1.1 Structure and properties of kombucha cellulose mats

The structure of the cellulose mats presents a highly crystalline network formed by a unique arrangement of ribbons. More specifically, the protofibrils of the glucan chains are formed outside the cytoplasm and then secreted out of the cell wall of the bacteria through the pores present on the surface. The fibrils are then arranged in a bundle to form a microfibril cellulose ribbon [29]. Interconnection between these ribbons leads to the formation of a network structure [48]. These microfibrils remain attached to the bacterial cell and later assemble to form thick fibrils called the macrofibrils [33].

This structural morphology endows the kombucha mats with distinct physical and chemical properties such as good chemical stability and ion exchange capacity, high surface area, high porosity, and high purity [3,9,21,52]. Due to their very porous network, they are characterised by high water holding capacity, retaining the water through hydrogen bonding [22]. The presence of surface hydroxyl groups allows modification of the cellulose pellicles with other materials [33], making them good scaffold materials for fuctionalization. Moreover, they present excellent mechanical properties, including superior mechanical strength and elasticity, when in a living state [33]. Biological properties include high biocompatibility due to their enhanced cell-matrix interactions, cell signalling pathways and ability to maintain cellular homeostasis [45], biodegradability, and ability to self-grow, self-repair and regenerate. In addition to the above physical, chemical, mechanical and biological characteristics, in our previous studies, we have demonstrated that the kombucha cellulosic mats show dynamics of electrical activity represented by spikes of electrical potential [2]. This spiking activity may be attributed to the waves of depolarisation travelling in the mats, emerging from metabolically triggered release of potassium [1].

Several factors can influence the above characteristics and allow tuning of their properties and programmable behaviours. Firstly, the type of microbes that are present in the mats which depends on various parameters such as the source, substrate provided for fermentation, metabolites produced during fermentation and the climate and geography of cultivation [43]. The time of fermentation is another important factor as it affects the yeast and bacterial cell count, increasing up to 14 days of fermentation and then starting to decrease [10]. Other factors include the concentration and type of tea and sugar used for the preparation of the culture broth, the temperature, oxygen availability and pH [27,34,60].

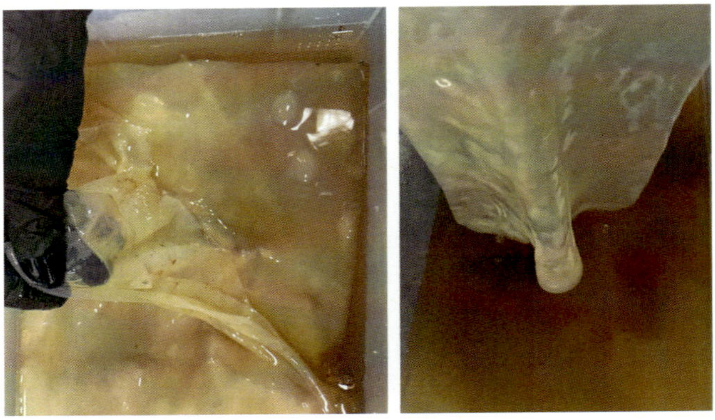

FIGURE 1
Kombucha mat formation in sugared tea infusion via microbial fermentation.

The above attributes make kombucha cellulose mats excellent sustainable scaffold matrices for the development of materials with responsive capabilities.

2 METHODS

2.1 Preparation of kombucha mats

For the preparation of the kombucha mats, a commercially obtained SCOBY acquired from Freshly Fermented Ltd (Lee-on-the-Solent, PO13 9FU, UK), was cultured in an infusion of 5 L of boiled tap water, 500 g of white granulated sugar (Tate & Lyle, UK) and 10 black tea bags (Taylors Yorkshire Teabags 125 g, UK) in a plastic container. The container was fully enclosed with a lid which had 8×0.5 mm holes to allow access to oxygen during the fermentation process, and stored in static conditions at room temperature of 19-21C. Every 10 days, the culture broth was replaced for the continuation of the fermentation and the production of nascent kombucha mats. When the mats were ready (4 weeks time), we proceeded with exposure to mechanical, electrical and optical stimulation(Figure 1).

2.2 Mechanical stimulation

For mechanical stimulation, load application of manually cut aluminium weights of 100 g (W51 L51 H14.2 mm), 200 g (W51 L51 H28.4 mm) and 300 g (W51 L51 H42.7 mm) was employed. In this study, unmodified and modified kombucha zoogleal mats were deployed for the evaluation of electrical responsive behaviours to compression. Graphene and zeolite nanoparticles were selected for the modification of the kombucha mats via in situ fermentation. Graphene exhibits excellent electrical conductivity and mechanical flexibility [55,61] while cellulose strongly binds with carbon materials [7, 20, 44, 62], making graphene a good candidate for enhancing the electrical properties of the cellulosic mats. Zeolites are crystalline aluminosilicates [4] with a large surface area, high adsorption capacity and high selectivity [5, 32, 36, 40], enabling the absorption of various components and facilitating the immobilisation and attachment of the graphene particles and cellulosic cells.

A nascent kombucha mat was functionalised with a mixture of 5.372g of graphene powder (HCG10000-P-1, Graphitene Energy Storage Series https://www.graphitene.com), 1.289g of Zeolite Y powder (ThermoFisher Scientific UK, 045862.36) and PTFE solvent (Sigma-Aldrich, 665880) to achieve good interfacial bonding between the graphene and zeolite molecules. The mixture was deposited on the surface of the mat using a razor blade and dispersed homogeneously until fully coated. After coating, the mat was folded to prevent spillage of the mixture and placed back into the culture solution. After 14 days of fermentation, a new 240x240 mm kombucha mat was formed, fully encapsulating the functionalised mat.

The modified kombucha mat and a nascent pristine mat were placed on top of a cardboard box each (W122 L160 mm) and secured in place with clamps. Both mats were subjected to weight application to attain loads of 100 g, 300 g and 400 g. Electrical activity recordings were performed for the duration of weight stimulation using an ADC-24 (Pico Technology, UK) high-resolution data logger with a 24-bit A/D converter. 8 pairs of platinum-iridium coated sub-dermal needle electrodes (Spes Medica Srl, thickness of 0.1 mm) were inserted into the mats, each pair reporting a difference of the electrical potential between the electrodes. The first 4 pairs (Channels 1-2, 3-4, 5-6, 7-8) were inserted into the unmodified mat and the remaining 4 (Channels 9-10, 11-12, 13-14, 15-16) into the modified mat. During the recordings, the logger undertook as many measurements as possible (typically up to 600 per second) and saved the average value(Figure 2). The distance between electrodes was 10-20 mm. All measurements were conducted at room temperature and the experiments were repeated to ensure error elimination.

Before analysing the spontaneous spiking activity of the loaded and unloaded mats, we comment on the passive conduction properties of the studied materials.

In dry conditions, the functionalized mat exhibits a resistance that is an order of magnitude lower than the pristine Kombucha mat due to the enhancements from the addition of graphene. When the mat is loaded consecutively, the resistance decreases, likely due to pseudoelastic stretching of the matrix under the normal forces of the loads, which reduces the available conduction section. Loading the functionalized mat results in higher currents (lower

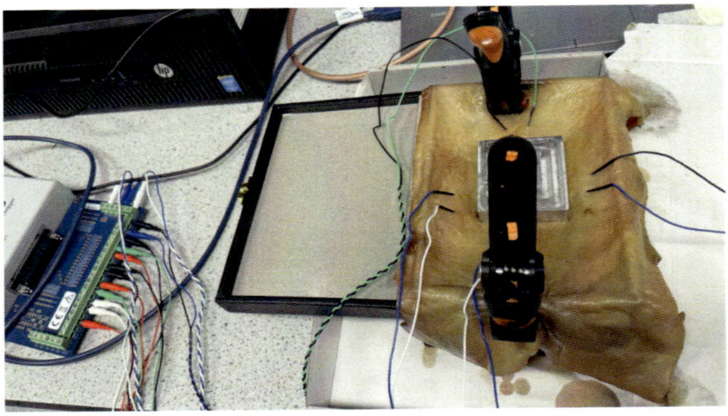

FIGURE 2
Electrical activity recordings of unmodified kombucha mat subjected to 100g square load. Four pairs of stainless steel sub-dermal needle electrodes were inserted into the sample to record the difference of the electrical potential between the electrodes (Channels 1-2, 3-4, 5-6, 7-8).

resistances), whereas loading the pristine mat results in lower currents (higher resistances). This effect can be attributed to the improved conductivity from the percolation of fillers added to the functionalized mat. Increased pressure creates more contact points conductive fillers, such as graphene flakes, thus reducing the mats resistance.

The spontaneous spiking activity of two mats was continuously recorded during load application. The profiles from four differential channels are presented in Figure 3 over an 11-day measurement period. Each channel displays a certain degree of correlation with transitions and increased spiking activity having remarkably similar characteristics. Typically, the range of spontaneous activity is smaller in a pristine mat, in the order of 1 mV, and higher in the functionalized mat. Moving to Fast Fourier Transform (FFT) and Higher Order Complexity (HOC) metrics [12–14] we can derive other useful information, as shown in Figure4.

The FFT analysis reveals how the noise spectrum across the mat sensors ranges from approximately 1 mHz up to 0.5 Hz, which represents our limits considering the sampling frequency (1 Hz). The entire figure is based on "Region 2", corresponding to the loading with a 400g load.

The red lines (differential electrode pairs 1-2, 3-4, 5-6, and 7-8) represent the pristine Kombucha, while the blue lines (differential electrode pairs 9-10, 11-12, 13-14, and 15-16) represent the functionalized Kombucha. A fundamental oscillation mode occurs across all channels at 1mHz, with visible harmonics at 2, 3, and 4 mHz. It is unclear whether this ultra-low frequency mode is due to external noise disturbance or part of the mat's electrical oscillations. Notably, there is a broad peak spanning 0.1 to 0.2 Hz, which is a

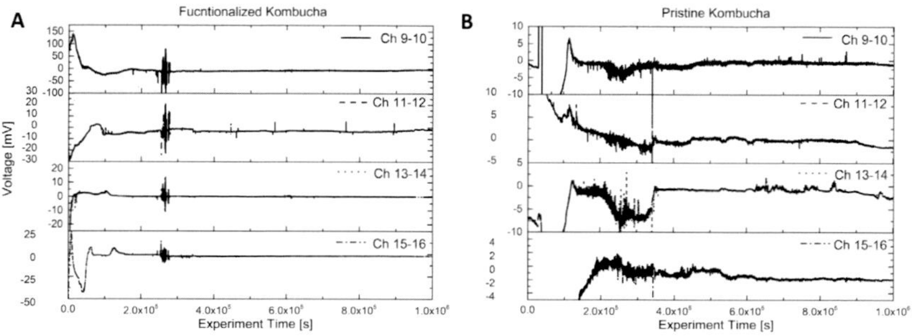

FIGURE 3
Recordings showing the spiking response of the mats. (A) Functionalised Kombucha response under loading and (B) pristine Kombucha response.

genuine signal component from the functionalized mat not observed in the pristine mat data.

HOC measures (panels C and D) undergo a depression (SE, TE, RE, SF, KE, LZC, DI) or a potentiation (SI, PCI, FD) caused by the load after 75 to 100 minutes, and lasting for approximately 100 minutes, in the same "Region 2". This is followed by a recovery to initial higher (lower) values. The functionalized Kombucha mat exhibits a much higher range of HOC measures explored, in comparison to the pristine mat. This indicates that proper implementation of HOC measures can interpret the mat's spontaneous oscillations as a sensor response over long timescales.

Further analysis using Short Time Fourier Transforms (STFT) generates spectrograms with colour-coded intensity fluctuations over time and frequency, depicted in Figure 5. Both panels show a portion of the measures exhibiting multiple harmonics and a complex structure resembling a vocal signal with fundamentals around 30 mHz and harmonics approaching 200 mHz. Such a complex structure is composed by quasi-parallel bands, seven of them can be easily counted, starting from the strongest (fundamental at 30 mHz, red band) and extending up to the highest harmonic (200 mHz, cyan band that is four times less strong than the fundamental). Such observation marks the complex behaviour of living organisms.

2.3 Electrical stimulation

For electrical stimulation, we subjected the kombucha mats to various sinusoidal input signals. We investigated potential correlations between input and output and therefore, the ability of the mats to transmit and process the electrical information. Thermal proteinoids and microalgae culture were used for the immobilisation of the kombucha pellicles. Proteinoids exhibit memristive, memcapacitive and conductive properties [41]. The proteinoids

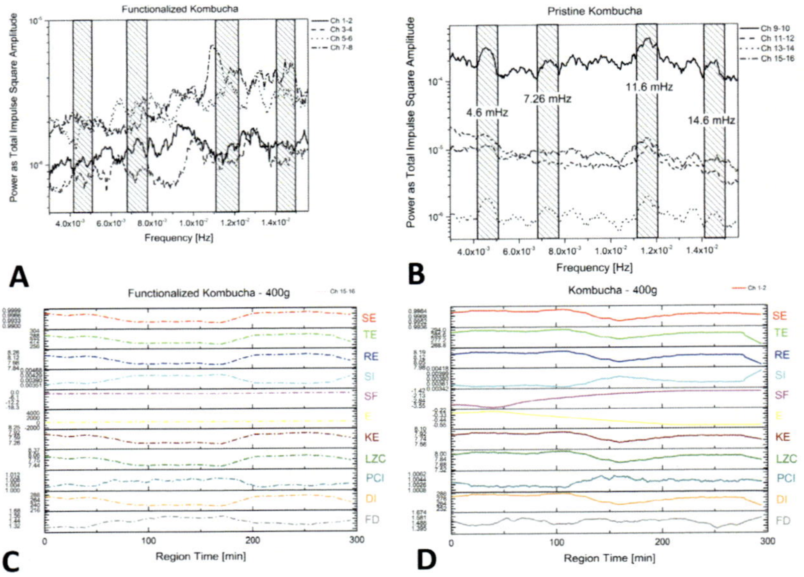

FIGURE 4
Regional details of the spiking activity features. (A) The Fast Fourier Transforms corresponds to 400 g loading of the functionalised Kombucha mat (A) and of the pristine mat (B). A zoomed-in view highlights a broad peak structure. (C) Higher-order complexity measures for "Region 2", functinoalized mat. SE stands for Shannon entropy, TE represents Tsallis entropy, RE stands for Rényi entropy, SI represents Simpson Index, SF represents Space Filling, E represents Expressiveness, KE represents Kolmogorov complexity, LZC represents Lempel-Ziv complexity, PCI represents Permuted Complexity Index, DI represents Diversity Index, and FD represents Fractal Dimension. (D) Higher-order complexity measures for "Region 2", pristine mat.

used were synthesised by heating the amino acids L-glutamic and L-arginine (Sigma Aldrich, purity level over 99%) using an existing thermal condensation protocol [42]. The microalgae *Chlorella vulgaris* was selected due to its robustness to changing environmental conditions, high growth rate and the presence of mutually beneficial interaction mechanisms with bacteria which allows the microalgae and the bacteria components of the kombucha mat to coexist [25, 50]. Moreover, we have previously reported that *C. vulgaris* exhibits periodic electroactivity in the form of spikes [46]. Combining the biological and electrical properties of the kombucha pellicles, proteinoids and microalgae opens up exciting possibilities for the development of a material with enhanced electron transport, conductivity and electrical properties. A nascent kombucha mat was placed in a 12 cm Petri dish with 20 ml kombucha infusion and a 12 cm Petri dish with 20 ml kombucha infusion and 10 ml *C. vulgaris* liquid culture (Blades Biological Ltd) to allow

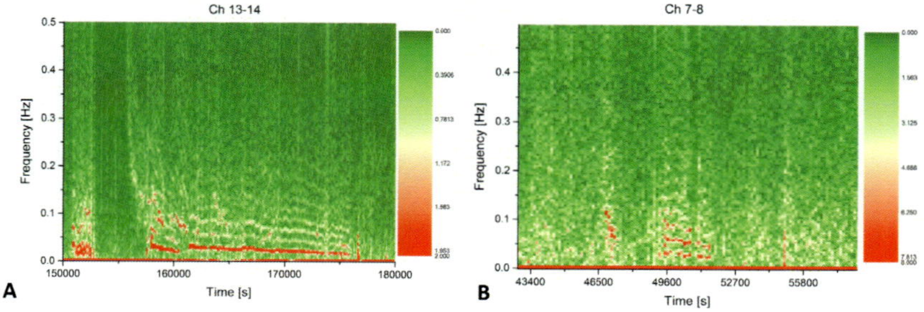

FIGURE 5
Regional details of the spiking activity features. (A) Short-time Fast Fourier Transforms on the region corresponding to loading of the differential channel 13-14 of the functionalized Kombucha mat. (B) Short-time Fast Fourier Transforms on the region corresponding to loading of the differential channel 7-8 of the pristine Kombucha mat.

further growth via fermentation. After a week, 2ml proteinoid solution were injected into the kombucha and kombucha-chlorella mats and placed back in the Petri dishes for 24h. Electrical stimulation was applied to the unmodified kombucha, kombucha -proteinoids and kombucha-proteinoids-chlorella mats. Platinum-iridium coated sub-dermal wire electrodes (Spes Medica Srl) were inserted into the mats at a distance of 10 mm from each other. BK Precision 4053B 10MHz dual channel function generator was used to supply input waveforms and the measurements were conducted at ambient temperature.

2.4 Optical stimulation

For optical stimulation, we subjected a nascent unmodified kombucha mat to pulses of yellow light at the frequencies of 2 Hz, 20Hz and 40 Hz to assess the impact of different frequencies of controlled light stimulation. Pairs of platinum-iridium coated electrodes (Spes Medica Srl, thickness of 0.1 mm) were inserted into the mat at a distance of 10 mm from each other. A Photonics F3000 system was used for the generation of the yellow pulses focusing on the kombucha mat sample. A high-definition 24-bit ADC-24 data logger (Pico Technology, UK) was used for the recordings of the electrical activity under light and no light exposure.

3 RESULTS

Distinct electrical responses of unmodified and modified kombucha mats to load application

The electrical activity recordings demonstrated that the kombucha mat presents distinct oscillatory activity when exposed to different weights. More

Channel	Peak potential	100 g load	300 g load	400 g load	Channel	Peak potential	100 g load	300 g load	400 g load
1-2	HPP (mV)	149.45	9.46	5.20	9-10	HPP (mV)	54.97	10.78	186.53
	LPP (mV)	3.47	4.41	1.07		LPP (mV)	4.57	1.00	4.00
3-4	HPP (mV)	6.18	3.04	3.84	11-12	HPP (mV)	8.07	13.41	45.88
	LPP (mV)	1.13	1.01	1.30		LPP (mV)	1.07	1.03	1.22
5-6	HPP (mV)	14.63	3.48	12.06	13-14	HPP (mV)	13.98	2.66	25.37
	LPP (mV)	1.01	1.00	1.59		LPP (mV)	1.02	1.01	1.01
7-8	HPP (mV)	41.36	2.18	3.26	15-16	HPP (mV)	29.56	2.09	21.69
	LPP (mV)	4.41	1.02	1.29		LPP (mV)	3.28	1.00	1.05

TABLE 1
Highest and lowest peak potentials across all channels under mechanical stimulation of 100 g, 300 g and 400 g load.

specifically, when subjected to 100 g load, the oscillatory activity ranged from the lowest peak potential of 1.01 mV to a highest peak potential of 149.45 mV, observed 9 hours from the start of the recordings. Exposure to 300 g weight demonstrated lower peak potential fluctuations reaching up to 9.46 mV after 147 hours. Stimulation of the kombucha mat with 400 g load reached a highest peak potential of 12.06 mV and lowest peak potential of 1.07 mV. It is worth noting that an initial rise and drop of potential is observed across all channels in the first few seconds and is attributed to load placement or replacement during the recordings.

Exposure of the functionalised kombucha mat to 100 g load presented a highest peak potential of 54.97, 68 hours from the beginning of the recording with lowest peak potential of 1.02 mV. When subjected to 300 g load, a highest peak potential of 13.41mV is reported, 62 hours from the beginning of the recording with lowest peak potential of 1 mV. Finally, exposure to 400 g load generated a distinct spiking activity response with spikes up to 186.53 mV, approximately 3.5 hours from the start of the recording. Figure 6 presents the potential (mV) vs time (sec) plots of the kombucha and functionalised kombucha mats under the different load application for the channels with highest peak potential values. The above results demonstrate that the functionalised kombucha mat exhibits broader potential fluctuations and higher sensitivity to external input compared to the kombucha mat, with the application of heavier loads (300 g and 400g) instigating greater potential response. Table 1 presents the highest and lowest peak potentials across all channels under mechanical stimulation of 100 g, 300 g and 400 g load. Similarly to the kombucha mat, an initial potential rise and drop is observed in the first few seconds due to load placement or replacement on the mat and is not taken into consideration in the values presented in the table.

A similar trend is observed when looking at the amplitude statistics. 400 g load application generated higher amplitude values for the functionalised kombucha mat compared to the unmodified kombucha mat, with Channel

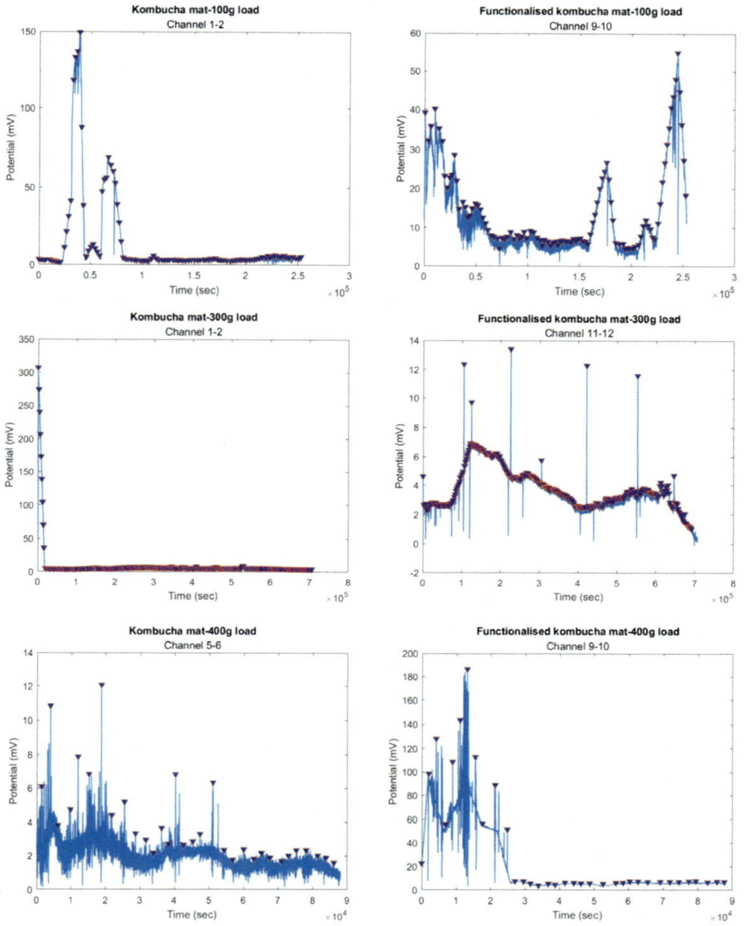

FIGURE 6
Plots depicting the potential (mV) vs time (sec) of the unmodified and modified with graphene and zeolite kombucha mats under 100 g, 300 g and 400 g loading for the channels with highest peak potential values.

9-10 exhibiting the highest mean value of 31.86 mV and standard deviation amplitude value of 49.31 mV. The highest median and mean amplitude values for the unmodified kombucha mat were observed in Channel 5-6 and were 2.63 mV and 3.75 mV accordingly, with a standard deviation value of 2.64 mV. Exposure to 100 g and 300 g weight produced variable amplitude values across all channels for both mats although 100 g load generated mean amplitude values significantly higher than the 300 g loading of both mats. The highest mean spike amplitudes were observed in Channel 1-2 (15.6 mV,

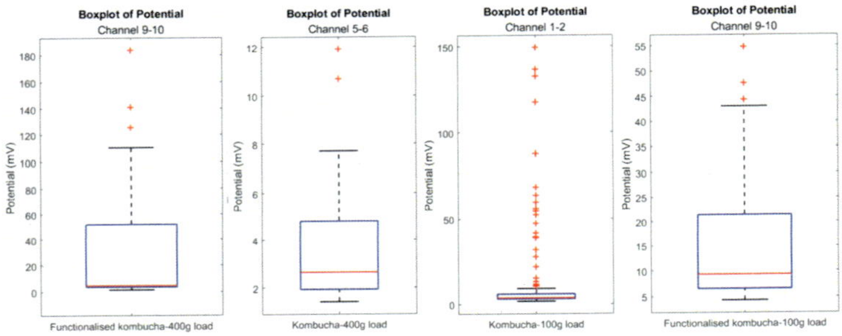

FIGURE 7
Distribution of amplitudes for the channels with highest spike amplitudes of the unmodified and modified kombucha mats. The median amplitude values are marked with the horizontal red line inside the box and the outliers shown as individual points represent the values beyond the minimum and maximum range. Channel 9-10 of the functionalised kombucha mat, subjected to 400 g load, presents significantly higher amplitude values.

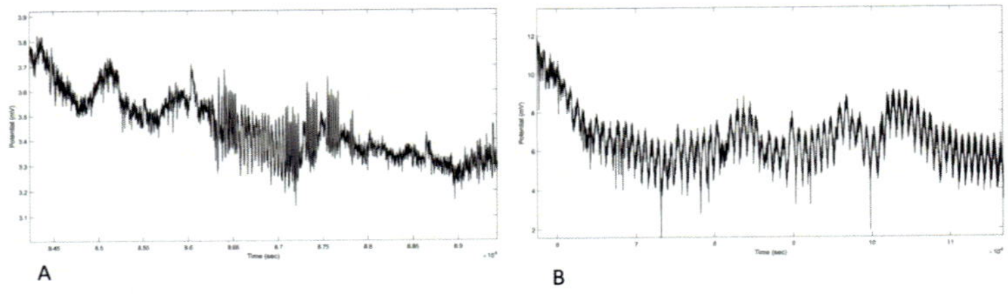

FIGURE 8
A.Response of unmodified kombucha mat to 100 g load. Example of low frequency trains of spikes across Channel 1-2 with a duration of 14 min., 1.07 min and 1.91 min. B. Response of modified kombucha mat to 100 g load. Example of low frequency trains of spikes reported across Channel 15-16.

$\sigma = 29.07$ and Channel 9-10 (14.87 mV, $\sigma = 11.62$). Figure 7 shows the distribution of amplitudes for the channels with highest spike amplitudes of the modified and unmodified kombucha mats.

Low frequency trains of spikes are observed for both the unmodified and modified kombucha mats when subjected to the different loads, with the functionalised mat generating higher amplitude trains of spiking activity(Figure 8).

The average distance between the spikes for the unmodified mat is recorded as below:

- For 100 g load application, the average distance recorded across the channels spans between 39.82 min (Channel 1-2, $\sigma = 8.83$) and 47.97 min (Channel 3-4, $\sigma = 29.57$).

- For 300 g load application, the average distance spans between 44.42 min (Channel 1-2, $\sigma = 10.08$) and 81.75 min (Channel 5-6, $\sigma = 174.97$).
- For 400 g load application, the average distance is recorded from 41.63 min (Channel 7-8 $\sigma = 7.77$) to 46.53 min (Channel 3-4, $\sigma = 44.4$).

While the functionalised kombucha mat exhibited an average distance between spikes:

- For 100 g load application, the average distance spans between 40.20 min (Channel 15-16, $\sigma = 8.83$) and 42.21 min (Channel 13-14, $\sigma == 8.93$).
- For 300 g load application, the average distance starts from 42.22 min (Channel 11-12, $\sigma = 8.65$) to 163.19 min (Channel 15-16, $\sigma = 285.13$).
- For 400 g load application, the average distance is recorded from 41.75 min (Channel 9-10 $\sigma = 11.90$) to 84.25 min (Channel 13-14, $\sigma = 108.82$).

100 g load application shows more consistent average spike duration values across the channels for both mats. A similar trend is observed for 400 g loading of the unmodified kombucha mat. However, 300 g loading for both mats and 400 g loading for the modified mat shows a wider range of average duration across the channels, with some of them presenting significantly increased average spike duration, indicating an increase of the frequency of the spikes. The larger potential fluctuations exhibited by the functionalised kombucha mat when subjected to heavier loads may be attributed to the addition of the graphene-zeolite nanoparticles. Since graphene exhibits enhanced conductivity, its presence in the kombucha mat could lead to the modification and potential amplification of the electrical activity signals. Mechanical stress could be induced with the weight application, causing alterations on the metabolic activity of the microbial cells. Additionally, the pressure from the load application could impact the oxygen availability and distribution in the mat, leading to distinct electrical activity.

Kombucha-Proteinoid Biofilms Exhibit Enhanced Signal Transduction and Tunable Response Characteristics

Proteinoid-based composites that include kombucha and chlorella showed enhanced ability to transmit signals compared to proteinoids alone when exposed to sinusoidal input signals. The kombucha-proteinoid biofilms displayed higher output amplitudes and stable frequencies. The samples that contained both kombucha and chlorella showed the greatest amplification. Conversely, the system consisting solely of proteinoids exhibited a non-linear correlation between the input and output signals. The amplitudes of the signals reduced as the input levels increased, indicating the presence of inherent filtering properties that rely on the amplitude. The increased signalling behaviour observed in kombucha-proteinoid composites is most likely due to the synergistic interactions between the different components. Hydrophilic polysaccharides and cell surfaces in kombucha and chlorella can hold more proteinoids due to electrostatic and hydrogen bonding interactions, which leads to increased conductivity. The presence of cellulosic and microbial nanowires in kombucha may enable the establishment of direct linkages between proteinoid nanostructures, so forming interconnected channels for the transfer of electrons. In addition, the endogenous redox proteins and metabolites of chlorella have the potential to interact with proteinoids, resulting in the formation of unique intramolecular charge transfer pathways. The proteinoids, kombucha, and chlorella produce a well-structured 3D matrix that improves the connectivity between neighbouring domains, facilitating effective and precise signal transmission. The interaction of kombucha and chlorella with proteinoids has the potential to cause changes in their structure, which could lead to the emergence of different conductivity states that have enhanced transmission efficiency. The collaborative molecular interactions among the components seem to improve the movement of electrons, transfer of charge, and ability to conduct electricity, leading to the effective transmission of amplified input signals.

These findings indicate that the combination of proteinoids and microbial components, such as kombucha, alters the signalling characteristics, resulting in enhanced qualities such as greater amplification and a wider range of frequency response. By adjusting the input-output relationships, it is possible to optimise the composition and attain the required response characteristics. Additional spectroscopic and microscopic research are required to clarify the exact structure-function mechanisms that are responsible for these interactions in the integrated biomolecular composites.

The signal transduction properties of proteinoid-kombucha biofilms were investigated using various compositions (Figure 9). The output amplitude (A_{out}) was found to be significantly higher in samples containing kombucha-chlorella (samples A and C) compared to those containing only kombucha (samples B and D), as described by the following equation:

$$A_{out} \propto \alpha \cdot C_{Chlorella} \tag{1}$$

where α is a proportionality constant and $C_{Chlorella}$ is the concentration of chlorella in the biofilm. The input-output relationships of the biofilms were

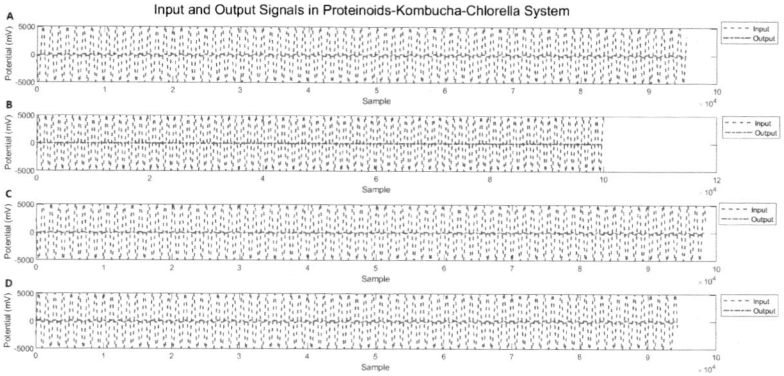

FIGURE 9
Composition-dependent signal transduction in proteinoid-kombucha biofilms. Samples A and C with chlorella have a much higher output potential than samples B and D with only kombucha. The input frequencies stay the same, and there are only small changes in the output frequencies.

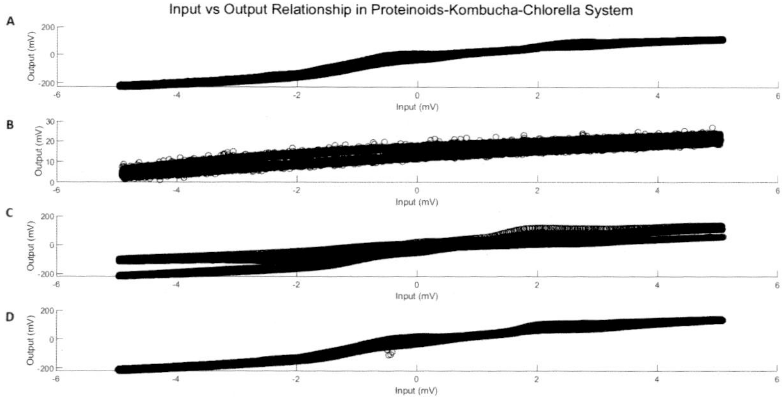

FIGURE 10
The graph shows the linear relationship between the input and output potentials for four different proteinoid-kombucha biofilm samples. The slopes of the lines vary depending on whether chlorella is present or absent in the biofilms. Given the same input potentials, samples A and C, which contain kombucha-chlorella, display steeper slopes, indicating their ability to generate higher output potentials compared to samples B and D, which only contain kombucha. This shows that proteinoid-kombucha biofilms can be tuned to send signals more efficiently in a variety of situations, with the effects changing in size depending on the input amplitude.

further analyzed (Figure 10), revealing a linear correlation between input potential (V_{in}) and output potential (V_{out}), with the slope (m) dependent on the biofilm composition:

$$V_{out} = m \cdot V_{in} + b \tag{2}$$

Parameter	Sample A	Sample B	Sample C	Sample D
Input Amplitude (mV)	10.044256	9.963069	10.055854	10.055853
Input Frequency (Hz)	0.055043	0.055313	0.055259	0.055047
Output Amplitude (mV)	352.276800	26.126230	360.593900	362.623600
Output Frequency (Hz)	0.113672	0.255910	0.119684	0.093523

TABLE 2
Input and output characteristics of proteinoid-kombucha biofilms with varying compositions.

where b is the y-intercept. The slopes were found to be steeper for kombucha-chlorella samples (A and C) compared to kombucha-only samples (B and D), indicating the tunability of the biofilms for optimised signal transmission.

The input and output characteristics of proteinoid-kombucha biofilms with varying compositions are presented in Table 2. While samples A and B have comparable input amplitudes and frequencies, their output properties differ considerably. Sample A, which includes kombucha-chlorella, exhibits a much greater output amplitude (352.276800 mV) in comparison to Sample B (26.126230 mV), which solely contains kombucha. The output frequencies differ between the two samples, with Sample A having a lower output frequency of 0.113672 Hz compared to Sample B's frequency of 0.255910 Hz.Both Samples C and D, which contain kombucha chlorella, exhibit uniform input amplitudes and frequencies. Sample A exhibits significantly larger output amplitudes (361.608750 mV) than Sample B, although its output frequencies (0.106604 Hz) are lower than both Samples A and B. The results indicate that the composition of biofilm has a substantial impact on signal transduction qualities. Specifically, the presence of chlorella leads to higher output amplitudes and lower output frequencies.

Quasi-Chaotic Signal Propagation in Kombucha-Based Living Films Stimulated with Logistic Map Inputs

The logistic map is a nonlinear difference equation that can exhibit complex, chaotic behaviour despite its simple form. The equation for the logistic map is given by:

$$x_{n+1} = r \cdot x_n \cdot (1 - x_n) \tag{3}$$

where x_n represents the population ratio at time step n, and r is a parameter that controls the behaviour of the system. For $0 \leq r \leq 4$, the population ratio remains bounded between 0 and 1. To investigate the quasi-chaotic signal propagation in kombucha-based living films, the logistic map is employed to generate input signals. The input voltage, V_{in}, is derived from the population ratio x_n as follows:

$$V_{in} = A \cdot (2x_n - 1) \tag{4}$$

where A is the amplitude scaling factor, and the term $(2x_n - 1)$ is used to center the input voltage around zero. The kombucha-based living films are stimulated with these quasi-chaotic input signals, and their response is measured as an output voltage, V_{out}. The relationship between the input and output signals can be characterized by a transfer function, $H(f)$, in the frequency domain:

$$V_{out}(f) = H(f) \cdot V_{in}(f) \tag{5}$$

where f represents the frequency. In this subsection, we present the results of stimulating kombucha-based living films with quasi-chaotic input signals generated using the logistic map equation. We analyse the input-output signal characteristics, including amplitude and frequency content, to comprehend the signal propagation properties of the living films under chaotic stimuli. We estimate the transfer function, $H(f)$, to gain insights into the frequency-dependent response of the films.

To convert the chaotic fractions generated by the logistic map into voltage signals, we applied a linear scaling process, adjusting the values to a range of -500 mV to +500 mV:

$$V_{in} = 1000 \cdot x_n - 500 \tag{6}$$

Rather than focusing on individual waveforms, we adopted an input-output analytic approach that compares statistical features of the system's response. This integration of mathematically constructed randomness and biophysical stimulation protocols allows for control of emergent bio-electronic systems. By adjusting the μ parameter, we can control the response of the chemical reaction network within the kombucha gate. Low values of μ lead to consistent, fixed concentration outputs, while high values of μ result in complex, fluctuating signals. The use of feedback loops to redirect output states back into the system enables the development of autonomous, self-contained chaotic circuits.To quantify the level of chaos in the system, we calculated the Lyapunov exponent for different increments of μ. When significantly deviating from homeostasis, positive Lyapunov exponents confirm the presence of deterministic chaos and highlight the intricate reorganisation of microenvironments. In our experiments, we used the logistic map to create voltage signals that were close to chaotic and used them to stimulate the kombucha networks. We analysed the input-output relationships by comparing statistical features such as mean, variance, and power spectral density. The Lyapunov exponent was calculated for various values of μ to quantify the level of chaos in the system's response. Moreover, we explored the impact of adding feedback loops, which redirected the output signals back into the system, enabling the dynamic modification of growth parameters. This enabled the development of autonomous, self-contained chaotic circuits within the kombucha networks.

The input voltage signals were applied to the kombucha networks, and the resulting output voltage responses were recorded for Channels C, D, E, and F. Figure 12A shows the voltage signal produced by the logistic map chaotic oscillator as the input, whereas Figure 12B illustrates the voltage responses of the kombucha networks over time. The box plots exhibit the average, median (50th percentile), 25th and 75th percentiles (edges of the box), and the standard deviation (whiskers) for each channel. The input signal (Channel A) has a mean of 13.56 mV and a standard deviation of 42.70 mV. The output channels (C-F) have increasing mean values and greater standard deviations. The mean values range from 9.58 mV to 58.14 mV, while the standard deviations range from 265.37 mV to 409.79 mV, respectively. The increase in box sizes and whisker lengths from Channel C to Channel F indicates the amplification and greater unpredictability of the output signals in response to the chaotic input, implying the existence of complex, non-linear dynamics inside the stimulated kombucha networks. The results indicate that kombucha networks has the capacity to effectively process and convert complex, disorderly signals. This could potentially result in the development of unique, self-organised behaviours in response to non-linear inputs. The observed amplification and filtering effects in the output channels indicate that kombucha networks have inherent signal processing capabilities that can be used for diverse applications in sensing, computation, and adaptive materials.

We studied the time-varying transfer functions of Channels C, D, E, and F in order to understand more about the signal processing capacities of kombucha networks in response to chaotic input. The chaotic input signal produced by the logistic map oscillator is processed and transformed by the kombucha networks as measured by the transfer functions, which are computed as the frequency domain ratio of the output voltage to input voltage. The transfer functions of the kombucha networks stimulated with chaotic input are shown over time in Figure 13. As the input signal moves across the network, it is gradually amplified as seen by the rise in mean transfer function values from Channel C (4.63) to Channel F (8.70). The transfer functions' standard deviations drop from 38.44 at Channel C to 18.58 at Channel F, indicating a decrease in unpredictability and perhaps a filtering effect.

Mutual Information Analysis Reveals Inter-Channel Dependencies in kombucha Networks

In order to examine the interconnections and data exchange between the output channels of the kombucha networks, we conducted a mutual information analysis on the transfer functions of Channels C, D, E, and F. Mutual information measures the extent of information that is common to two random

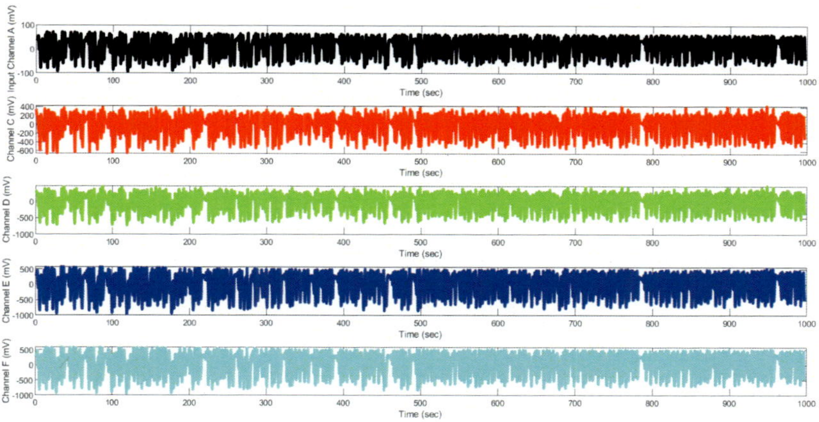

FIGURE 11
A) The logistic map chaotic oscillator generates an input voltage signal, whereas (B) the kombucha networks (Channels C, D, E, and F) produce output voltage responses when stimulated by the chaotic input signal over time. The input signal displays fast, chaotic variations that are characteristic of chaotic dynamics. The output channels exhibit amplified and filtered representations of the input signal, characterised by variable levels of complexity and non-linearity. The transfer functions, determined by dividing the output by the input in the frequency domain, offer a numerical evaluation of the signal processing functions of the kombucha networks. The average transfer function values show a rise from Channel C (4.63) to Channel F (8.70), suggesting a gradual increase in the amplification of the input signal. The standard deviations of the transfer functions fall as the signal propagates from Channel C (38.44) to Channel F (18.58), indicating a decrease in uncertainty and a potential filtering impact within the network. The results indicate that kombucha networks have the capacity to process and convert complex, disorderly signals, which could result in the emergence of new, self-organised behaviours in response to non-linear inputs.

variables, offering valuable understanding of the interactions and dependencies between the channels.

Figure 14 displays a heatmap illustrating the mutual information values between every pair of output channels. The diagonal portions of the heatmap indicate the self-information or entropy of each channel, while the off-diagonal elements display the mutual information across distinct channels. The self-information (entropy) of each channel was calculated using the following equation:

$$H(X) = -\sum_{x \in X} p(x) \log_2 p(x) \tag{7}$$

where $H(X)$ is the entropy of channel X, and $p(x)$ is the probability of observing value x in the transfer function of channel X. The mutual

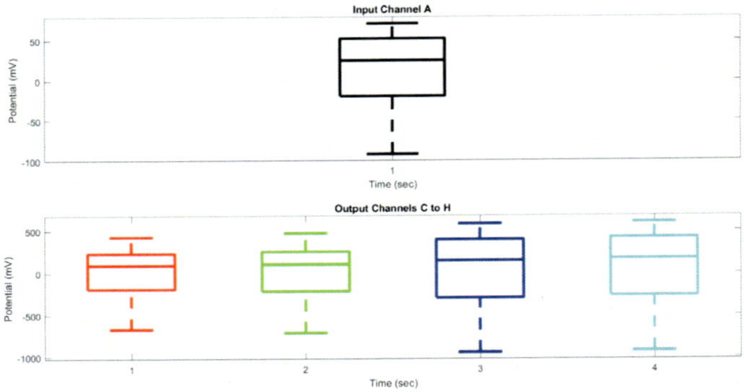

FIGURE 12
Box graphs illustrating the statistical characteristics of (A) input voltage signals produced by the logistic map chaotic oscillator and (B) output voltage responses of kombucha networks triggered by the chaotic input. The box plots illustrate the average, the middle value (50th quantile), the lower and upper quartiles (shown by the edges of the box), and the variability (represented by the whiskers) for each channel. The input signal (Channel A) has a mean of 13.56 mV and a standard deviation of 42.70 mV. The output channels (C-F) have increasing mean values and larger standard deviations. The mean values range from 9.58 mV to 58.14 mV, and the standard deviations range from 265.37 mV to 409.79 mV, respectively. The increasing dimensions of the boxes and lengths of the whiskers from Channel C to Channel F indicate the amplification and greater unpredictability of the output signals in response to the chaotic input. This suggests the existence of complex, non-linear dynamics inside the stimulated kombucha networks.

	Channel C	Channel D	Channel E	Channel F
Channel C	0.5499	0.3520	0.2835	0.3418
Channel D	–	1.1688	0.3245	0.3004
Channel E	–	–	0.7882	0.4364
Channel F	–	–	–	1.3727

TABLE 3
Mutual information values for output channels C, D, E, and F of the kombucha networks.

information between two channels, X and Y, was calculated using:

$$MI(X;Y) = \sum_{x \in X} \sum_{y \in Y} p(x,y) \log_2 \frac{p(x,y)}{p(x)p(y)} \tag{8}$$

where $MI(X;Y)$ is the mutual information between channels X and Y, $p(x,y)$ is the joint probability distribution of X and Y, and $p(x)$ and $p(y)$ are the marginal probability distributions of X and Y, respectively.

The mutual information values for each channel are presented in Table 3. Channel F has a self-information (entropy) value of 1.3727, which indicates a greater level of complexity and unpredictability in its transfer function.

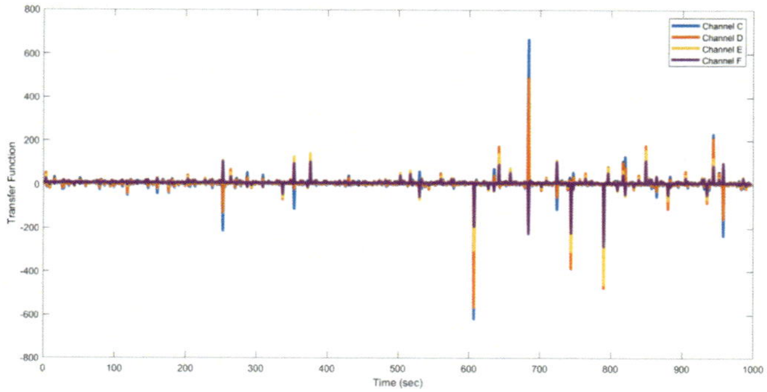

FIGURE 13
Transfer functions of kombucha networks (Channels C, D, E, and F) over time when stimulated with a chaotic input generated by the logistic map oscillator. The transfer functions, obtained by dividing the output voltage by the input voltage in the frequency domain, quantify the signal processing capacities of the kombucha networks. The average transfer function values show a rise from Channel C (4.63) to Channel F (8.70), suggesting a gradual increase in the amplification of the input signal as it travels through the network. The standard deviations of the transfer functions fall as we move from Channel C (with a standard deviation of 38.44) to Channel F (with a standard deviation of 18.58). This indicates a decrease in variability and suggests the presence of a probable filtering effect. The varied transfer functions of the kombucha networks demonstrate their dynamic and adaptive responsiveness to chaotic input. The presence of clearly defined peaks and troughs in the transfer functions indicates the existence of resonant frequencies and complex, non-linear interactions within the stimulated kombucha networks. The results illustrate the capacity of kombucha networks to analyse and convert disorderly signals, which could potentially result in the development of new, self-organised behaviours in response to non-linear inputs.

Channel D has the highest level of information sharing with other channels, as indicated by mutual information values of 0.3520, 0.3245, and 0.3004 with Channels C, E, and F, respectively. Channel D appears to have an essential role in the transmission of information and interconnections within the kombucha networks. The non-symmetrical characteristics of the mutual information values, as depicted in Figure 14 and Table 3, emphasise the existence of directed information transmission and complex interconnections among the output channels. As an illustration, the mutual information between Channels C and D ($MI_{CD} = 0.3520$) is greater than the mutual information between Channels D and C ($MI_{DC} = 0.3520$), suggesting that Channel C has a greater effect on Channel D compared to the reverse. The results indicate the presence of complex, non-linear connections and exchange of information among the output channels of the kombucha networks when exposed to a chaotic input. The analysis of mutual information indicates the existence of directed information flow and asymmetrical dependencies, which may enhance the kombucha networks' capacity to process and react to chaotic inputs.

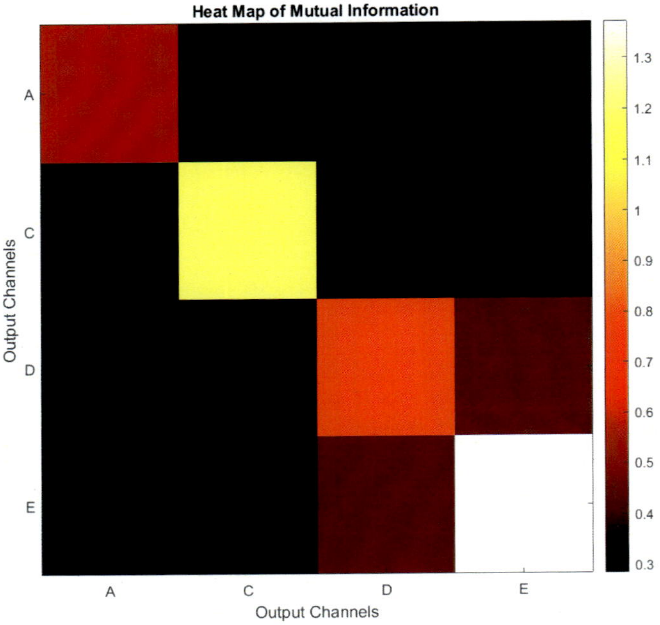

FIGURE 14
A heatmap displaying the mutual information values between the transfer functions of output channels C, D, E, and F of the kombucha networks when stimulated with a chaotic input created by the logistic map oscillator. The diagonal elements reflect the entropy of each channel, whereas the off-diagonal elements indicate the mutual information across distinct channels. Higher mutual information values indicate more pronounced interactions and information exchange between channels. The heatmap displays the presence of differences among the channels, where Channel F demonstrates the highest level of self-information and Channel D has the greatest quantity of information shared with other channels. The results indicate the existence of complex, non-linear connections and information transmission within the kombucha networks, which could enhance their capacity to process and respond to chaotic inputs.

3.1 Optical stimulation

The present research examined the electrical response of kombucha to yellow light stimulation at various frequencies, including no light, 2 Hz, 20 Hz, and 40 Hz. Figure 15a illustrates box plots that compare the spike amplitudes (mV) in each condition. In comparison to the no-light condition, the results indicate that light stimulation increases spike amplitudes, with the highest amplitudes observed at 2 Hz (median: 20.33 mV). The median spike amplitudes (medians: 18.64 mV and 16.11 mV, respectively) reduce slightly as the stimulation frequency increases to 20 Hz and 40 Hz. Table 4 confirms the frequency-dependent effects of yellow light stimulation on kombucha's electrical activity by providing extensive descriptive statistics for the spike amplitudes. The box diagrams in Figure 15b illustrate the periods between

Parameter	No Light	2 Hz	20 Hz	40 Hz
Amplitude (mV)				
Quartiles	4.66, 6.86, 8.77	16.37, 20.33, 27.28	15.84, 18.64, 22.04	11.66, 16.11, 21.95
Mean	7.01	22.28	19.81	18.25
Max	14.25	50.19	48.87	46.74
Min	1.89	6.19	3.94	2.16
Standard Deviation	3.31	8.46	6.82	8.94
Period (sec)				
Quartiles	673.00, 751.50, 874.00	706.25, 813.00, 1012.50	704.00, 805.00, 1013.25	698.00, 788.50, 969.50
Mean	818.55	874.47	873.08	856.82
Max	1357.00	1339.00	1351.00	1351.00
Min	78.00	78.00	116.00	186.00
Standard Deviation	213.80	193.52	196.92	192.25

TABLE 4
Comparison of spike amplitudes and periods in kombucha under different yellow light stimulation frequencies.

spikes (sec) across various light stimulation frequencies. The median periods are relatively consistent across all conditions, with a modest increase in comparison to the no-light condition (751.50 sec). This implies that yellow light stimulation has a less pronounced effect on the temporal dynamics of kombucha's electrical activity, despite the fact that it affects spike amplitudes in a frequency-dependent manner. In an effort to enhance the understanding of the intrinsic electrical properties of kombucha, Figure 16a illustrates the electrical potential (mV) over time (sec) in the absence of light. The graph indicates oscillatory activity that is irregular in nature, with periods (mean: 818.55 sec, SD: 213.80 sec) and amplitudes (mean: 7.01 mV, SD: 3.31 mV) that vary. A power-law decay and a dominant low-frequency component are evident in the power spectral density (PSD) analysis of Figure 16b, suggesting the presence of long-range correlations and slow oscillations in the electrical activity. Figure 16c illustrates the temporal evolution of the dominant frequency components through the continuous wavelet transform (CWT) scalogram. This reveals the intermittent occurrence of higher frequency fluctuations and the stability of weak oscillations. These findings indicate that kombucha shows intrinsic oscillatory electrical activity, which can be influenced by external light stimulation in a frequency-dependent manner. The results indicate that kombucha's electrical properties make it a promising responsive material for a variety of applications.

This study aimed to examine the electrical reaction of kombucha to yellow light stimulation at various frequencies (2 Hz, 20 Hz, and 40 Hz) and compare it to the baseline electrical activity when no light was present. The results of our study indicate that kombucha displays inherent rhythmic electrical behaviour, which is influenced by external light stimuli in a manner that depends on the frequency. In the absence of light, the electrical potential of kombucha exhibited unpredictable variations with an average amplitude of

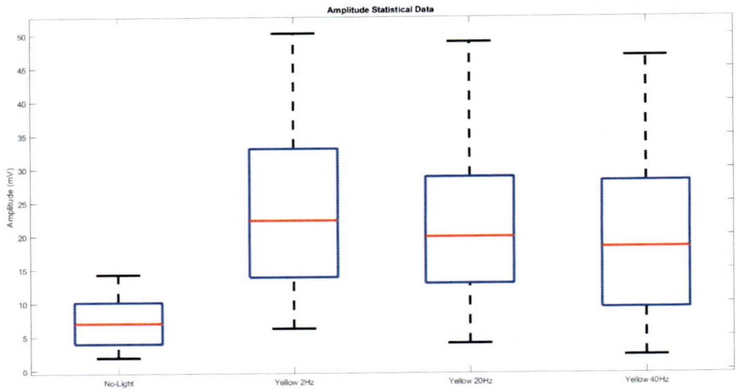

(a)

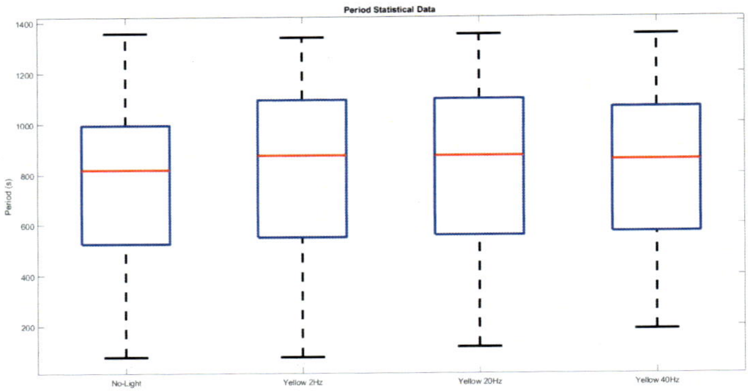

(b)

FIGURE 15
a) Box plots comparing the spike amplitudes (mV) of kombucha's electrical response under different yellow light stimulation frequencies (no light, 2 Hz, 20 Hz, and 40 Hz). The median spike amplitudes are 6.86 mV (no light), 20.33 mV (2 Hz), 18.64 mV (20 Hz), and 16.11 mV (40 Hz). The box plots show that light stimulation increases spike amplitudes compared to the no-light condition, with the highest amplitudes observed at 2 Hz. As the stimulation frequency increases (20 Hz and 40 Hz), the spike amplitudes decrease slightly. The whiskers and outliers indicate the range and variability of the spike amplitudes under each condition. This comparison highlights the frequency-dependent effects of yellow light stimulation on kombucha's electrical activity, with lower frequencies producing stronger responses. b) Box plots comparing the periods between spikes (sec) of kombucha's electrical response under different yellow light stimulation frequencies (no light, 2 Hz, 20 Hz, and 40 Hz). The median periods are 751.50 sec (no light), 813.00 sec (2 Hz), 805.00 sec (20 Hz), and 788.50 sec (40 Hz). The box plots show that the periods between spikes remain relatively similar across all light stimulation frequencies, with a slight increase compared to the no-light condition. The whiskers and outliers indicate the range and variability of the periods under each condition. This comparison suggests that while yellow light stimulation affects the spike amplitudes in a frequency-dependent manner, it has a less pronounced effect on the temporal dynamics of kombucha's electrical activity.

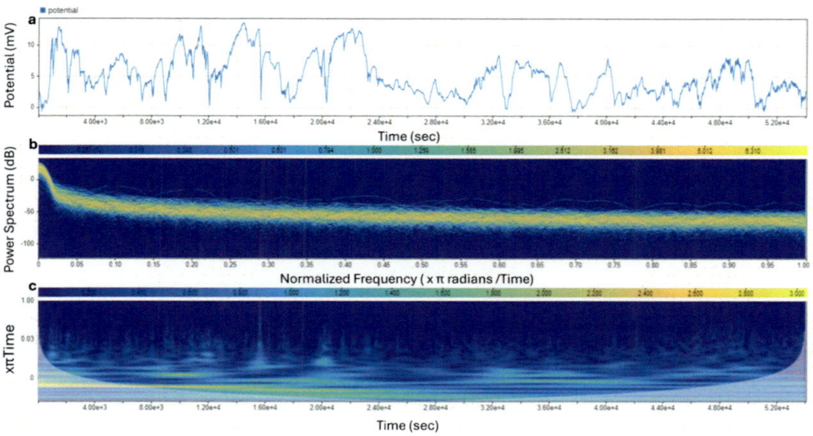

FIGURE 16
a) Electrical potential (mV) vs time (sec) of kombucha in the absence of light. The graph shows variations in electrical potential over time, with a mean amplitude of 7.01 mV (SD = 3.31 mV, min = 1.89 mV, max = 14.25 mV, quartiles: 4.66, 6.86, 8.77 mV). The mean period between peaks is 818.55 sec (SD = 213.80 sec, min = 78.00 sec, max = 1357.00 sec, quartiles: 673.00, 751.50, 874.00 sec). These findings suggest that kombucha exhibits oscillatory electrical activity even in the absence of light stimuli. The time series plot reveals the presence of irregular fluctuations in the electrical potential, with varying amplitudes and periods. The oscillatory behaviour appears to be intrinsic to kombucha's bioelectrical properties, as it persists in the absence of external light stimulation. b) Power spectral density (PSD) of kombucha's electrical potential in the absence of light. The persistent spectrum reveals the frequency components present in the oscillatory electrical activity. The PSD shows a dominant low-frequency component, indicating the presence of slow oscillations in the electrical potential. The spectrum also displays a power-law decay, suggesting the presence of long-range correlations in the electrical activity. This analysis provides insights into the frequency domain characteristics of kombucha's endogenous electrical oscillations. c) Continuous wavelet transform (CWT) scalogram of kombucha's electrical potential in the absence of light. The scalogram represents the time-frequency distribution of the electrical oscillations, with warmer colors indicating higher wavelet coefficients. The scalogram reveals the temporal evolution of the dominant frequency components in the electrical activity. The presence of persistent low-frequency components across time suggests the stability of the slow oscillations in kombucha's electrical potential. The scalogram also highlights the intermittent occurrence of higher frequency components, indicating the presence of transient fluctuations in the electrical activity. This time-frequency analysis complements the insights gained from the time series and power spectral density plots.

7.01 mV and an average time interval between peaks of 818.55 sec (Table 4). The presence of baseline oscillatory behaviour indicates that the microbial community of kombucha possesses inherent bioelectrical capacities that produce these electrical signals. The existence of this inherent electrical activity in kombucha aligns with other research that has documented bioelectrical events in microbial systems [28, 47]. Stimulation with yellow light at all investigated frequencies (2 Hz, 20 Hz, and 40 Hz) resulted in a significant increase in spike amplitudes relative to the absence of light (Table 4). The average spike amplitudes during light stimulation varied between 18.25 mV and 22.28 mV, indicating a two- to three-fold rise relative to the initial measurement. This discovery suggests that the microbial population of kombucha is responsive to external light stimulation and exhibits increased electrical activity as a result. Photons are believed to interact with microbial cells, causing an increase in spike amplitudes. This interaction may affect the cells' metabolic activities and ion exchange mechanisms, resulting in enhanced bioelectrical signalling. Remarkably, the electrical reaction of kombucha to light stimulation displayed a pattern that varied depending on the frequency. The lowest stimulation frequency (2 Hz) resulted in the highest mean spike amplitude (22.28 mV), although slightly lower amplitudes (19.81 mV and 18.25 mV, respectively) were detected at higher frequencies (20 Hz and 40 Hz). The fact that the reaction of kombucha's microbial population is affected by the frequency of light stimulation indicates that there is a specific range of frequencies that leads to the highest level of electrical activity. The reduced reaction at higher frequencies may be attributed to the barriers in the microbial cells' capacity to process and react to swift light pulses [11]. Unlike the variations in spike amplitudes, the intervals between spikes remained reasonably consistent across all frequencies of light stimulation, with average periods ranging from 856.82 seconds to 874.47 seconds (Table 4). The recorded values were slightly greater than the average duration in the absence of light (818.55 sec), indicating that exposure to light may have a minor impact on the temporal patterns of kombucha's electrical activity. Nevertheless, the variations in time intervals among the light stimulation conditions were not as noticeable as the variances reported for the spike amplitudes.

4 DISCUSSION

The results of this work have significant implications for understanding the bioelectrical characteristics of kombucha and its prospective uses as a material that can respond to external stimuli. The inherent bioelectrical capabilities of kombucha are demonstrated by its intrinsic rhythmic electrical activity, which occurs even without external stimulus. The responsiveness of kombucha to light stimulation at various frequencies indicates its potential as a programmable biohybrid system, where the intended electrical output may be controlled by modifying the parameters of light stimulation [51]. Moreover, the fact that kombucha's electrical response to light stimulation varies depending on the frequency provides opportunities for the advancement of biosensors and bioelectronic devices based on kombucha. Through the use of the distinct reactions to different light wavelengths, kombucha can be manipulated to identify and convert particular visual inputs into electrical outputs [38]. This characteristic has the potential to be used in the fields of environmental monitoring, optogenetics, and biohybrid computing [53]. Similarly, the distinct responses of the kombucha hybrid material to mechanical stimulation may allow tailored electrical properties through the modification of load application. The study also examined the electrical response of kombucha to other stimuli, including as sinusoidal input signals and quasi-chaotic signals generated using the logistic map equation. The results demonstrate that living films made from kombucha have distinctive signal transduction characteristics that may be adjusted by modifying the film's composition. Kombucha-proteinoid biofilms exhibited superior signal transmission capacities when subjected to sinusoidal input signals, in comparison to proteinoids alone. Moreover, the work investigated the transmission of quasi-chaotic signals in living films based on kombucha, using inputs from the logistic map. The findings indicated that kombucha networks exhibited a high level of efficiency in processing and converting complex, disordered data. The output channels displayed amplified and filtered representations of the input signal, which were distinguished by different degrees of complexity and nonlinearity. The transfer functions of the kombucha networks, which measure the signal processing capabilities, demonstrated a progressive amplification of the input signal and a reduction in variability as the signal passed through the network. An examination of mutual information in the output channels indicated the existence of complex, nonlinear interactions and information sharing within the kombucha networks. The research revealed the presence of directed information flow and asymmetrical dependencies among the channels, which could enhance the kombucha networks' capacity to process and respond to chaotic inputs.

However, additional research is required to completely clarify the mechanisms that explain kombucha's electrical reaction to stimulation. Examining the function of particular microbial species, their metabolic pathways, and the impact of light and pressure on cellular activities may yield more profound understanding of the observed bioelectrical phenomena [39]. Furthermore, investigating the impact of different light characteristics, such as wavelength and intensity, on the electrical activity of kombucha could further improve our understanding of its sensitive features [18, 26, 49].

The responsiveness of kombucha to mechanical, electrical and optical stimulation demonstrates its promise as a versatile biohybrid material that may be used in several fields including as biosensing, bioelectronics, and biohybrid systems. Additional investigation into the fundamental mechanisms and examination of supplementary light, mechanical and electrical characteristics may reveal unexplored possibilities for using kombucha's distinctive bioelectrical capabilities in diverse scientific and technological domains. The incorporation of proteinoids, chlorella and graphene-zeolites in the biofilms led to distinct spiking activity that can be related to the synergistic interactions between the components, leading to the emergence of interconnected pathways for electron transfer and the generation of various conductivity states. The findings illustrate the potential of kombucha-based living films as biomaterials that can process and convert input signals in a responsive manner. Kombucha-based systems show promise as potential candidates for a range of applications, including biosensors, biohybrid devices, and adaptive materials, due to their ability to adjust film composition and manage complex, chaotic inputs, hence allowing for tuning of signal transduction capabilities. Eventually, the application of electrical, mechanical and optical stimulation to kombucha-derived living films demonstrates their capacity as adaptable biomaterials with adjustable signal transduction characteristics. The capacity to analyse and convert complex, disorderly signals emphasises the potential to develop adaptable biohybrid systems for diverse applications (Fig. 17). Additional investigation in this field has the potential to result in a development of cutting-edge biosensors, biohybrid technologies, and intelligent materials that utilise the distinct characteristics of kombucha and its microbial community.

5 CONCLUSION

Overall, this study showcases the potential of kombucha-based living films as biomaterials that can respond to stimuli and have adjustable signal transduction properties. The combination of Kombucha, proteinoids, and chlorella led to the formation of biofilms that exhibited improved signal transmission capacities, increased output amplitudes, and consistent frequencies when compared to biofilms composed only of proteinoids. The amplification effect seen in the kombucha-proteinoid biofilms is a result of the synergistic interactions between the components. These interactions include the generation of interconnecting pathways for electron transfer and the emergence of various conductivity states. Moreover, the study of quasi-chaotic signal propagation in living films made from kombucha, using inputs from the logistic map, demonstrated their capacity to regulate and convert complex, chaotic signals. The output channels displayed amplified and filtered versions of the input signal, which were distinguished by different degrees of complexity and non-linearity. The transfer functions measured the signal processing capacity of the kombucha networks, demonstrating a gradual rise in the amplification of the input signal and a decrease in variability as the signal passed through the network. An examination of mutual information revealed the existence of complex, non-linear connections and information sharing within

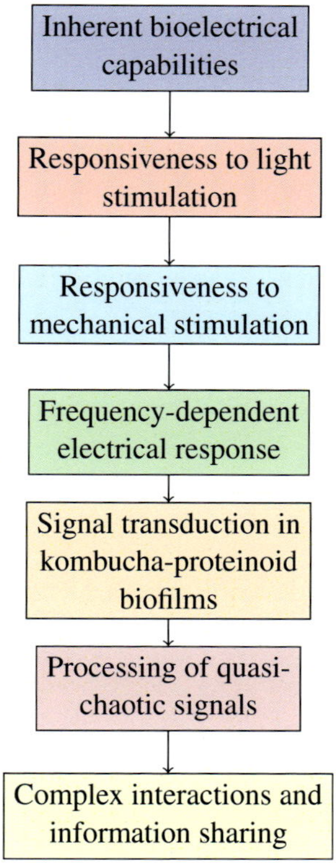

FIGURE 17
This diagram presents an overview of the key properties and characteristics of kombucha mats as responsive materials. These properties, which include inherent bioelectrical properties, responsiveness to light and mechanical stimulation, frequency-dependent electrical responses, enhanced signal transduction in kombucha-proteinoid biofilms, efficient processing of quasi-chaotic signals, and complex interactions and information sharing, contribute to the unique behaviour and potential applications of kombucha-based materials in various fields such as biosensing, bioelectronics, and biohybrid systems.

the kombucha networks. The controlled flow of information and varying connections among the channels enhance the kombucha networks' capacity to adapt and react to chaotic inputs. These findings create new opportunities for the advancement of adaptive biohybrid systems, biosensors, and smart materials that exploit the distinctive characteristics of kombucha and its microbial community. Additional research is necessary to have an in-depth knowledge of the fundamental mechanisms that account for the reported electrical characteristics of kombucha-based living films. Examining the function of particular types of microorganisms, their biochemical processes, and the impact of environmental factors on the electrical activity of kombucha could offer significant knowledge to develop customised biohybrid systems.

ACKNOWLEDGEMENTS

Authors are thankful to Niel Philips for his assistance with the instruments, David Patton for his contribution in SEM imaging, and Dr. Jun Yao for her invaluable advice on graphene and zeolites. The research on proteinoids was supported by EPSRC Grant EP/W010887/1, titled "Computing with Proteinoids. The data on the optical properties of kombucha and kombucha proteinoids, as well as chaos calculations, are available online and can be accessed at https://zenodo.org/records/12794215.

REFERENCES

[1] Andrew Adamatzky. (2022). Dynamics of electrical resistance of kombucha zoogleal mats. *Biophysical Reviews and Letters*, 17(04):135–144.

[2] Andrew Adamatzky. (2023). Electrical potential spiking of kombucha zoogleal mats: A symbiotic community of bacteria and yeasts. *Bioelectricity*, 5(2):99–108.

[3] Mădălina G Albu, Zina Vuluga, Denis M Panaitescu, Dumitru M Vuluga, Angela Căşărică, and Marius Ghiurea. (2014). Morphology and thermal stability of bacterial cellulose/collagen composites. *Central European Journal of Chemistry*, 12:968–975.

[4] Scott M Auerbach, Kathleen A Carrado, and Prabir K Dutta. (2003). *Handbook of zeolite science and technology*. CRC press.

[5] Noerma J Azhari, Nadya Nurdini, St Mardiana, Thalabul Ilmi, Adroit TN Fajar, IGBN Makertihartha, Grandprix TM Kadja, *et al.* (2022). Zeolite-based catalyst for direct conversion of co2 to c2+ hydrocarbon: A review. *Journal of CO2 Utilization*, 59:101969.

[6] Douglas A Balentine, Sheila A Wiseman, and Liesbeth CM Bouwens. (1997). The chemistry of tea flavonoids. *Critical Reviews in Food Science & Nutrition*, 37(8):693–704.

[7] Lihong Bao and Xiaodong Li. (2012). Towards textile energy storage from cotton t-shirts. *Advanced materials*, 24(24):3246–3252.

[8] Somnath Chakravorty, Semantee Bhattacharya, Antonis Chatzinotas, Writachit Chakraborty, Debanjana Bhattacharya, and Ratan Gachhui. (2016). Kombucha tea fermentation: Microbial and biochemical dynamics. *International journal of food microbiology*, 220:63–72.

[9] Prashant R Chawla, Ishwar B Bajaj, Shrikant A Survase, and Rekha S Singhal. (2009). Microbial cellulose: fermentative production and applications. *Food Technology & Biotechnology*, 47(2).

[10] C Chen and BY Liu. (2000). Changes in major components of tea fungus metabolites during prolonged fermentation. *Journal of applied microbiology*, 89(5):834–839.

[11] Adam E Cohen and Veena Venkatachalam. (2014). Bringing bioelectricity to light. *Annual review of biophysics*, 43:211–232.

[12] Mohammad Mahdi Dehshibi and Andrew Adamatzky. (2021). Electrical activity of fungi: Spikes detection and complexity analysis. *Biosystems*, 203:104373.

[13] Mohammad Mahdi Dehshibi and Andrew Adamatzky. (2023). *Fungal Machines: Sensing and Computing with Fungi*, chapter Complexity of Electrical Spiking of Fungi, pages 33–60. Springer Nature Switzerland, Publication Location.

[14] Mohammad Mahdi Dehshibi, Alessandro Chiolerio, Anna Nikolaidou, Richard Mayne, Antoni Gandia, Mona Ashtari-Majlan, and Andrew Adamatzky. (2021). Stimulating Fungi Pleurotus Ostreatus with Hydrocortisone. *ACS Biomaterials Science & Engineering*, 7(8):3718–3726.

[15] Andréia Boechat Delatorre, Tania Reis Lichtenfels Ribeiro, Nycole Mota Cortat Arcênio, and Geórgia Peixoto Bechara Mothé. (2020). Estudo sobre a adequação da normativa 41 para a produção e comercialização de kombucha: estudo de caso em uma empresa de macaé. In *Congresso de Ensino Pesquisa e Extensão-CONEPE.*

[16] Eirini Dimidi, Selina Rose Cox, Megan Rossi, and Kevin Whelan. (2019). Fermented foods: definitions and characteristics, impact on the gut microbiota and effects on gastrointestinal health and disease. *Nutrients*, 11(8):1806.

[17] C Dufresne and E Farnworth. (2000). Tea, kombucha, and health: a review. *Food research international*, 33(6):409–421.

[18] Ari Dwijayanti, Congqiang Zhang, Chueh Loo Poh, and Thomas Lautier. (2022). Toward multiplexed optogenetic circuits. *Frontiers in Bioengineering and Biotechnology*, 9:804563.

[19] G Frank. (1994). Kombucha, health beverage and natural remedy from the far east. steyr. *House Ennsthaler.*

[20] Kezheng Gao, Ziqiang Shao, Jia Li, Xi Wang, Xiaoqing Peng, Wenjun Wang, and Feijun Wang. (2013). Cellulose nanofiber–graphene all solid-state flexible supercapacitors. *Journal of Materials Chemistry A*, 1(1):63–67.

[21] G Gayathry and G Gopalaswamy. (2014). Production and characterisation of microbial cellulosic fibre from acetobacter xylinum.

[22] Kristina Gelin, Aase Bodin, Paul Gatenholm, Albert Mihranyan, Katarina Edwards, and Maria Strømme. (2007). Characterization of water in bacterial cellulose using dielectric spectroscopy and electron microscopy. *Polymer*, 48(26):7623–7631.

[23] WN Goh, A Rosma, B Kaur, A Fazilah, AA Karim, and Rajeev Bhat. (2012). Fermentation of black tea broth (kombucha): I. effects of sucrose concentration and fermentation time on the yield of microbial cellulose. *International Food Research Journal*, 19(1).

[24] WN Goh, A Rosma, B Kaur, A Fazilah, AA Karim, and Rajeev Bhat. (2012). Microstructure and physical properties of microbial cellulose produced during fermentation of black tea broth (kombucha). ii. *International Food Research Journal*, 19(1):153.

[25] Zhi Guo and Yen Wah Tong. (2014). The interactions between chlorella vulgaris and algal symbiotic bacteria under photoautotrophic and photoheterotrophic conditions. *Journal of Applied Phycology*, 26:1483–1492.

[26] Stephan Hirschi, Thomas R Ward, Wolfgang P Meier, Daniel J Müller, and Dimitrios Fotiadis. (2022). Synthetic biology: bottom-up assembly of molecular systems. *Chemical reviews*, 122(21):16294–16328.

[27] Weili Hu, Shiyan Chen, Jingxuan Yang, Zhe Li, and Huaping Wang. (2014). Functionalized bacterial cellulose derivatives and nanocomposites. *Carbohydrate polymers*, 101:1043–1060.

[28] Jacqueline Humphries, Liyang Xiong, Jintao Liu, Arthur Prindle, Fang Yuan, Heidi A Arjes, Lev Tsimring, and Gürol M Süel. (2017). Species-independent attraction to biofilms through electrical signaling. *Cell*, 168(1):200–209.

[29] M Iguchi, S Yamanaka, and A Budhiono. (2000). Bacterial cellulosea masterpiece of nature's arts. *Journal of materials science*, 35(2):261–270.

[30] Jill Jarrell, Tiffany Cal, and JW Bennett. (2000). The kombucha consortia of yeasts and bacteria. *Mycologist*, 14(4):166–170.

[31] Daria Kaczmarczyk, Stanisław Lochyński, *et al.* (2014). Products of biotransformation of tea infusion–properties and application. *Polish Journal of Natural Sciences*, 29(4):381–392.

[32] Grandprix TM Kadja, Mutiara D Rukmana, Rino R Mukti, M Haris Mahyuddin, Adhitya G Saputro, and Triati DK Wungu. (2021). Solvent-free, small organic lactam-assisted synthesis of zsm-5 zeolites. *Materials Letters*, 290:129501.

[33] D Laavanya, Shivanand Shirkole, and P Balasubramanian. (2021). Current challenges, applications and future perspectives of scoby cellulose of kombucha fermentation. *Journal of Cleaner Production*, 295:126454.

[34] Koon-Yang Lee and Alexander Bismarck. (2012). Susceptibility of never-dried and freeze-dried bacterial cellulose towards esterification with organic acid. *Cellulose*, 19:891–900.

[35] C-H Liu, W-H Hsu, F-L Lee, and C-C Liao. The isolation and identification of microbes from a fermented tea beverage, haipao, and their interactions during haipao fermentation. *Food microbiology*, 13(6).

[36] A Maghfirah, MM Ilmi, ATN Fajar, and GTM Kadja. (2020). A review on the green synthesis of hierarchically porous zeolite. *Materials Today Chemistry*, 17:100348.

[37] R Malbaša, E Lončar, and M Djurić. (2008). Comparison of the products of kombucha fermentation on sucrose and molasses. *Food chemistry*, 106(3):1039–1045.

[38] Sumit Malik, Joginder Singh, Rohit Goyat, Yajvinder Saharan, Vivek Chaudhry, Ahmad Umar, Ahmed A Ibrahim, Sheikh Akbar, Sadia Ameen, and Sotirios Baskoutas. (2023). Nanomaterials-based biosensor and their applications: A review. *Heliyon*.

[39] Roland Mathis and Martin Ackermann. (2016). Response of single bacterial cells to stress gives rise to complex history dependence at the population level. *Proceedings of the National Academy of Sciences*, 113(15):4224–4229.

[40] Svetlana Mintova, Jean-Pierre Gilson, and Valentin Valtchev. (2013). Advances in nano-sized zeolites. *Nanoscale*, 5(15):6693–6703.

[41] Panagiotis Mougkogiannis and Andrew Adamatzky. (2024). Memfractance of proteinoids. *ACS omega*, 9(13):15085–15100.

[42] Panagiotis Mougkogiannis, Neil Phillips, and Andrew Adamatzky. (2023). Transfer functions of proteinoid microspheres. *Biosystems*, 227:104892.

[43] Tehmeena Ashraf Mukadam, Kapil Punjabi, Sunita D Deshpande, Shashikant Prabhakar Vaidya, and Abhay Shadashiv Chowdhary. (2016). Isolation and characterization of bacteria and yeast from kombucha tea. *International Journal of Current Microbiology and Applied Sciences*, 5(6):32–41.

[44] Leif Nyholm, Gustav Nyström, Albert Mihranyan, and Maria Strømme. (2011). Toward flexible polymer and paper-based energy storage devices. *Advanced Materials*, 23(33):3751–3769.

[45] Sukanya Patra, Monika Singh, Divya Pareek, Kirti Wasnik, Prem S Gupta, and Pradip Paik. (2022). Advances in the development of biodegradable polymeric materials for biomedical applications.

[46] Neil Phillips, Richard Mayne, and Andrew Adamatzky. (2022). Chlorella sensors in liquid marbles and droplets. *Sensing and Bio-Sensing Research*, 36:100491.

[47] Arthur Prindle, Jintao Liu, Munehiro Asally, San Ly, Jordi Garcia-Ojalvo, and Gürol M. Süel. (2015). Ion channels enable electrical communication in bacterial communities. *nature*, 527(7576):59–63.

[48] Peter Ross, Raphael Mayer, and Moshe Benziman. (1991). Cellulose biosynthesis and function in bacteria. *Microbiological reviews*, 55(1):35–58.

[49] Andrey Rozenberg, Keiichi Inoue, Hideki Kandori, and Oded Béjà. (2021). Microbial rhodopsins: the last two decades. *Annual Review of Microbiology*, 75:427–447.

[50] Carl Safi, Bachar Zebib, Othmane Merah, Pierre-Yves Pontalier, and Carlos Vaca-Garcia. (2014). Morphology, composition, production, processing and applications of chlorella vulgaris: A review. *Renewable and Sustainable Energy Reviews*, 35:265–278.

[51] Mahmut Selman Sakar, Jeroen Eyckmans, Roel Pieters, Daniel Eberli, Bradley J Nelson, and Christopher S Chen. (2016). Cellular forces and matrix assembly coordinate fibrous tissue repair. *Nature communications*, 7(1):11036.

[52] Nasrullah Shah, Mazhar Ul-Islam, Waleed Ahmad Khattak, and Joong Kon Park. (2013). Overview of bacterial cellulose composites: a multipurpose advanced material. *Carbohydrate polymers*, 98(2):1585–1598.

[53] Lin Shao, Shi Luo, Zhihui Wang, Xinzhao Xu, Yongkun Yan, Yangjiang Wu, Meng Guo, Dacheng Wei, Yan Zhao, and Yunqi Liu. (2022). A flexible biohybrid reflex arc mimicking neurotransmitter transmission. *Cell Reports Physical Science*, 3(7).

[54] Martin Sievers, Cristina Lanini, Adrien Weber, Ursula Schuler-Schmid, and Michael Teuber. (1995). Microbiology and fermentation balance in a kombucha beverage obtained from a tea fungus fermentation. *Systematic and applied microbiology*, 18(4):590–594.

[55] Ji Won Suk, Richard D Piner, Jinho An, and Rodney S Ruoff. (2010). Mechanical properties of monolayer graphene oxide. *ACS nano*, 4(11):6557–6564.

[56] Ai Leng Teoh, Gillian Heard, and Julian Cox. (2004). Yeast ecology of kombucha fermentation. *International journal of food microbiology*, 95(2):119–126.

[57] Selorm Torgbo and Prakit Sukyai. (2018). Bacterial cellulose-based scaffold materials for bone tissue engineering. *Applied Materials Today*, 11:34–49.

[58] Thierry Tran, Cosette Grandvalet, François Verdier, Antoine Martin, Hervé Alexandre, and Raphaëlle Tourdot-Maréchal. (2020). Microbial dynamics between yeasts and acetic acid bacteria in kombucha: Impacts on the chemical composition of the beverage. *Foods*, 9(7):963.

[59] Thierry Tran, François Verdier, Antoine Martin, Hervé Alexandre, Cosette Grandvalet, and Raphaëlle Tourdot-Maréchal. (2022). Oxygen management during kombucha production: Roles of the matrix, microbial activity, and process parameters. *Food Microbiology*, 105:104024.

[60] Mazhar Ul-Islam, Waleed Ahmad Khattak, Minkyung Kang, Sang Min Kim, Taous Khan, and Joong Kon Park. (2013). Effect of post-synthetic processing conditions on structural variations and applications of bacterial cellulose. *Cellulose*, 20:253–263.

[61] Guoxiu Wang, Juan Yang, Jinsoo Park, Xinglong Gou, Bei Wang, Hao Liu, and Jane Yao. (2008). Facile synthesis and characterization of graphene nanosheets. *The Journal of Physical Chemistry C*, 112(22):8192–8195.

[62] Xing Xie, Guihua Yu, Nian Liu, Zhenan Bao, Craig S Criddle, and Yi Cui. (2012). Graphene–sponges as high-performance low-cost anodes for microbial fuel cells. *Energy & Environmental Science*, 5(5):6862–6866.

Complexity and Nonlinearity of Colloid Electrical Transducers

RAPHAEL FORTULAN[1,*], NOUSHIN RAEISI KHEIRABADI[1], ALESSANDRO CHIOLERIO[1,2] AND ANDREW ADAMATZKY[1]

[1] *Unconventional Computing Laboratory, UWE, Bristol, UK*
[2] *Bioinspired Soft Robotics Laboratory, Istituto Italiano di Tecnologia, Via Morego 30, 16165 Genova, Italy*

Received: September 1, 2024. Accepted: September 28, 2024.

This work explores the complexity and nonlinearity of seven different colloidal suspensions—Au, ferrofluid, TiO_2, ZnO, g–C_3N_4, MXene, and PEDOT:PSS—when electrically stimulated with fractal, chaotic, and random binary signals. The recorded electrical responses were analyzed using entropy, file compression, fractal dimension, and Fisher information measures to quantify complexity. The nonlinearity introduced by each colloid was evaluated by the deviation of the output from the best-fit hyperplane of the input-output mapping. The results showed that TiO_2 was the most complex colloid across all inputs, exhibiting high entropy, poor compressibility, and an unpredictable response pattern. The colloids also exhibited significant nonlinearity, making them promising candidates for reservoir computation, where the mapping of inputs into high-dimensional nonlinear states is advantageous. This study provides insight into the dynamics of colloids and their potential for unconventional computational applications that exploit their inherent complexity and nonlinearity, and it provides a rapid method for assessing the suitability of a particular material for use as a computational substrate before others.

Keywords: Colloids, complexity analysis, nonlinearity, unconventional computing

1 INTRODUCTION

Colloids, dispersed nanoparticles in a liquid, exhibit rich dynamics and physical phenomena arising from the interactions between the particles and the suspending medium. The application of external fields, such as electric or

* Corresponding author: E-mail: raphael.vicentefortulan@uwe.ac.uk

magnetic fields, can further influence the behavior of these systems, leading to the emergence of complex patterns and nonlinear responses. Harnessing these dynamics holds promise for unconventional computing paradigms, including reservoir and *in-materia* computing [21, 24–26, 39].

The notion of computing with liquid media, also known as liquid computing, dates back over 120 years to early hydraulic algebraic machines [1]. Since then, various approaches like fluid maze solvers, droplet logics, and colloid-based information processors have been explored, taking advantage of unique properties such as reconfigurability, scalability, potential low energy consumption, and inherent parallelism [13, 36]. In addition, previous works have demonstrated that colloid systems can implement synaptic learning and pattern recognition [8, 9].

In this work, we explore the complexity and nonlinearity of seven distinct colloidal suspensions—Au, ferrofluid, TiO_2, ZnO, g–C_3N_4, MXene, and PEDOT:PSS—when subjected to fractal, chaotic, and random binary electrical signals. By analyzing the recorded electrical responses using various complexity measures and quantifying the degree of nonlinearity introduced by the colloids, we aim to identify the most promising colloidal systems for potential use in reservoir computing architectures. The findings from this study provide valuable insights into the rich dynamical behaviors exhibited by colloids under electrical stimulation and their suitability for unconventional computing applications.

2 MATERIALS FABRICATION

A total of seven colloids were prepared to study their complexity and nonlinearity. Each colloid preparation method is described as follows:

Au: A colloid solution of Au nanoparticles dispersed in H_2O with a particle size of 14 nm and a concentration of 0.1 mg ml^{-1} was provided by PlasmaChem GmbH (Germany).

Ferrofluid: A ferrofluid solution containing Fe_3O_4 nanoparticles dispersed in H_2O at a concentration of 30 wt% was supplied by PlasmaChem GmbH (Germany).

TiO_2: A stable colloid of TiO_2 nanoparticles (30-50 nm), anatase phase, in H_2O 40 wt% was purchased from US Research Nanomaterials (USA).

ZnO: Sodium dodecyl sulfate (SDS, Merck) was added to deionized water to create a surfactant solution with a concentration of 0.22 wt%. The solution was then stirred continuously to homogenize. In sequence, 1 mg ZnO nanoparticles (US Research Nanomaterials, USA) were added to dimethyl sulfoxide (DMSO, US Research Nanomaterials, USA) while stirring continuously. The mixture was then stirred while 2 ml of SDS solution and 1 ml of $10\,\mathrm{mol\,l^{-1}}$ NaOH were added. The resultant dispersion concentration was maintained constant at $0.1\,\mathrm{mg\,ml^{-1}}$. The stirring operation was then continued for a few more hours to achieve a homogeneous dispersion of ZnO [20, 33].

g–C_3N_4: To synthesize g–C_3N_4 nanosheets, urea was first finely ground in an agate mortar and pestle. Then 25 g of the powder was placed in an alumina crucible, sealed, and heated up in a muffle furnace up to 550 °C at a rate of $25\,^{\circ}\mathrm{C\,min^{-1}}$. This temperature was maintained for 3 h to attain a yellow g–C_3N_4. The acquired g–C_3N_4 was then ground into powder using an agate mortar and pestle [32]. To achieve a suspension of g–C_3N_4 in H_2O with a concentration of $1\,\mathrm{mg\,ml^{-1}}$, 10 mg of g–C_3N_4 was mixed with 10 ml deionized water [22].

MXene: In the first stage of MXene synthesis, a sacrificial etching process was employed. A solution of lithium fluoride (LiF, 0.8 g) in hydrochloric acid (HCl, 10 ml, $9\,\mathrm{mol\,l^{-1}}$) was stirred for 5 min, facilitating the in situ production of a small amount of hydrofluoric acid (HF). The MAX phase powder (0.5 g Ti_3AlC_2) was then gradually added to the etching solution under continuous stirring. The sealed container was then placed on a magnetic stirrer at 35 °C and 1000 rpm for 24 h. This allowed selective etching of the aluminum layers within the MAX phase by the HF [2].

Following the etching step, the resultant mixture was centrifuged (3500 rpm, 5 min) with deionized water. This procedure eliminated the etching residue and excess acid. The centrifugation cycles were repeated until the supernatant reached a neutral pH ($\sim$6). According to reports, the remaining supernatant is a green, stable colloidal solution of MXene flakes dispersed in water at a low concentration ($1\,\mathrm{mg\,ml^{-1}}$), indicating that delamination has been achieved [15]. The remaining, darker part, which most likely contained larger MXene nanosheets, was sonicated with 50 ml of deionized water for 1 h to encourage further delamination. The sonicated mixture was then recentrifuged to separate the water. Sediment containing MXene flakes was collected and dried in a vacuum oven at 120 °C. The dried MXene powder was finally stored in a sealed container for future use [11, 16]. To prepare a suspension of MXene in dimethylformamide (DMF) at a concentration of $1\,\mathrm{mg\,ml^{-1}}$, 10 mg of as-synthesized MXene was mixed with 10 ml DMF under stirring 1000 rpm for 24 h.

PEDOT:PSS: A high-conductivity grade solution of PEDOT:PSS, with a concentration of 3-4 wt% in H_2O, was purchased from Merck. Before conducting the studies, the suspensions were subjected to an ultrasonic bath (40 kHz, DK Sonic Ultrasonic, UK) for 30 min to guarantee uniform dispersion of the particles.

3 EXPERIMENTAL SETUP

The electrical signals were generated using an Arduino Uno R4 board (Arduino, Italy) connected via I2C to two MCP4728 12-bit quad-channel digital-to-analog converters (DACs, Microchip Technology, USA), using a 74HC595 multiplexer (Texas Instruments, USA). The output channels of the DACs were connected to amplifier/level shift circuits based on LM258AP op-amps (Texas Instruments, USA) to generate bipolar outputs of ± 5 V. Both positive and negative voltage rails were supplied by an IPS 4303 Laboratory DC Power Supply (ISO-TECH, Taiwan). The outputs were then electrically connected to the colloids using $\varnothing 10\,\mu$m platinum/iridium-coated stainless steel probes to be used as inputs. An output probe was inserted into the colloid, and its voltage was measured using a 24-bit data logger from Pico Technology, UK.

To investigate the complexity and nonlinearity of all suspensions, three distinct signal patterns were programmed into the microcontroller: a random binary signal, a fractal curve, and a chaotic signal. The random binary signal, composed of a sequence of 1s and 0s generated randomly, aimed to probe the response of colloids to stochastic perturbations. The fractal curve, exhibiting self-similar patterns at different scales, was chosen to explore the hierarchical nature of colloidal structures. Meanwhile, the chaotic signal, characterized by its sensitive dependence on initial conditions, was intended to study the potential chaotic dynamics within the colloids.

By subjecting colloids to various stimuli, we aimed to understand their capabilities for advanced sensing and complex computation, as well as their usage in unconventional computing and reservoir computing. Each signal is described in more detail in the following sections.

3.1 Koch snowflake

Koch snowflake (or Koch curve) is one of the earliest and most famous fractal curves [41]. In this work, the curve was generated using Algorithm S1, as shown in the supporting information. The curve represented by two signals x and y was then applied to the colloid using two probes.

3.2 Lorenz system

The Lorenz system is a chaotic dynamical system and is modeled by the following set of differential equations

$$\dot{x} = \sigma(y - x), \tag{1}$$

$$\dot{y} = x(\rho - z), \tag{2}$$

$$\dot{z} = xy - \beta z, \tag{3}$$

where σ, ρ, and β are parameters. In this work, the dynamical system was implemented on the microcontroller using the 4th-order Runge-Kutta method [7] with a time step of 100 ms. The choice of parameters was $\sigma = 10$, $\rho = 28$, and $\beta = 8/3$, and the initial condition was set to $[1.2, 1.4, 1.6]^{\mathsf{T}}$.The x, y, and z states were then applied to the colloid using three probes.

3.3 Random Bombardment

For the random bombardment, a random number between 0 and 255 (which corresponds to 8 bits) was selected every 500 ms, converted to its binary representation, and applied to the material using 8 probes. Bipolar logic was used, with +5 V representing the logical 1 and -5 V representing the logical 0.

4 BLACK-BOX COMPLEXITY

We picture that the colloid is a black box that hides a function $f : V \to W$ inside. Assuming that no structural information is known about f, the only way to learn about it is by evaluating the function values of $f(x)$ at potential values $x \in V$. In addition, it is also possible to assume that the evaluation is performed by some *oracle*, from which $f(x)$ is queried, which represents the signal generator and acquisition system. The complexity of f (and by extension of the colloid) can then be inferred from a set of discrete tuples $(x_0, f(x_0)), (x_1, f(x_1)), \ldots, (x_{N-1}, f(x_{N-1}))$.

For chaotic and dynamical systemssystems characterized by sensitive dependence on initial conditions and nonlinear dynamicsthe definition of complexity for a time series can be split into two categories: (1) regularity and (2) predictability [6,38]. Regularity refers to the number of repetitive patterns in the signal or the smoothness of the signal. Predictability, on the other hand, refers to the correlation present in the signal, i.e., how much information it carries about itself. In this work, several complexity measures are used to characterize the output time series data of the colloid. Fractal dimension quantifies the complexity repeated in the signal as the ratio of the log number of boxes to the log of the inverse box length, $\log(N)/\log(1/L)$ [42]. Shannon's entropy, $H(X) = -\sum_{i=0}^{N-1} P(x_i)\log_2 P(x_i)$ [37], measures the minimum data storage requirements. Fisher information quantifies the amount of self-information carried by the system, which is anti-correlated with complexity. Compression algorithms like Gzip, bzip2, and LZMA2 assess compressibility, with more complex or random signals being less compressible. Further details on these complexity measures are available in the supporting information.

The computed entropy, gzip size, bzip size, XZ size, Fisher information, and fractal dimension of the recorded electrical signals from the colloids

Colloid	**Entropy**	**gzip Size**	**bzip2 Size**	**XZ Size**	**FI**	**FD**
Au	10.5	5.41	4.74	4.06	7.63E-01	1.02
Ferrofluid	10.5	3.53	2.99	2.71	9.00E-01	1.06
g–C_3N_4	10.5	5.25	4.65	4.13	6.25E-01	1.03
MXene	10.4	5.35	4.81	4.08	4.95E-01	1.02
PEDOT:PSS	10.5	5.35	4.78	4.08	7.50E-01	1.01
TiO_2	11.2	5.39	4.68	4.48	2.10E-01	1.05
ZnO	10.5	4.94	4.35	3.79	6.73E-01	1.03

TABLE 1
The calculated entropy, gzip size, bzip size, XZ size, Fisher information, and fractal dimension of the measured electrical signal of the colloids for a fractal-like signal input

Colloid	**Entropy**	**gzip Size**	**bzip2 Size**	**XZ Size**	**FI**	**FD**
Au	12.4	4.95	4.41	3.58	7.83E-01	1.02
Ferrofluid	12.4	3.44	2.88	2.62	9.47E-01	1.03
g–C_3N_4	12.3	4.58	3.99	3.29	6.32E-01	1.06
MXene	12.4	4.65	4.18	3.23	7.71E-01	1.00
PEDOT:PSS	12.4	5.10	4.43	3.80	8.77E-01	1.04
TiO_2	12.4	5.16	4.35	3.99	5.60E-01	1.07
ZnO	18.5	4.43	3.80	3.03	5.36E-01	1.02

TABLE 2
The calculated entropy, gzip size, bzip size, XZ size, Fisher information, and fractal dimension of the measured electrical signal of the colloids for a chaotic-like signal input

Colloid	**Entropy**	**gzip Size**	**bzip2 Size**	**XZ Size**	**FI**	**FD**
Au	12.6	5.92	5.05	5.32	6.16E-02	1.08
Ferrofluid	11.3	4.18	3.36	3.50	9.11E-01	1.08
g–C_3N_4	20.5	5.90	5.07	5.44	1.53E-03	1.09
MXene	16.1	5.76	4.93	5.21	7.95E-03	1.11
PEDOT:PSS	12.1	5.88	5.08	5.23	1.04E-01	1.08
TiO_2	38.0	5.87	5.05	5.34	1.23E-03	1.10
ZnO	15.9	5.68	4.86	5.11	9.75E-03	1.09

TABLE 3
The calculated entropy, gzip size, bzip size, XZ size, Fisher information, and fractal dimension of the measured electrical signal of the colloids for an 8-bit random binary signal input

under fractal, chaotic, and random stimuli are presented in Tabs. 1 to 3, respectively. The most complex colloidal suspension was chosen as the one exhibiting the highest complexity indices and was identified as the most complex. Subsequent selections proceeded accordingly.

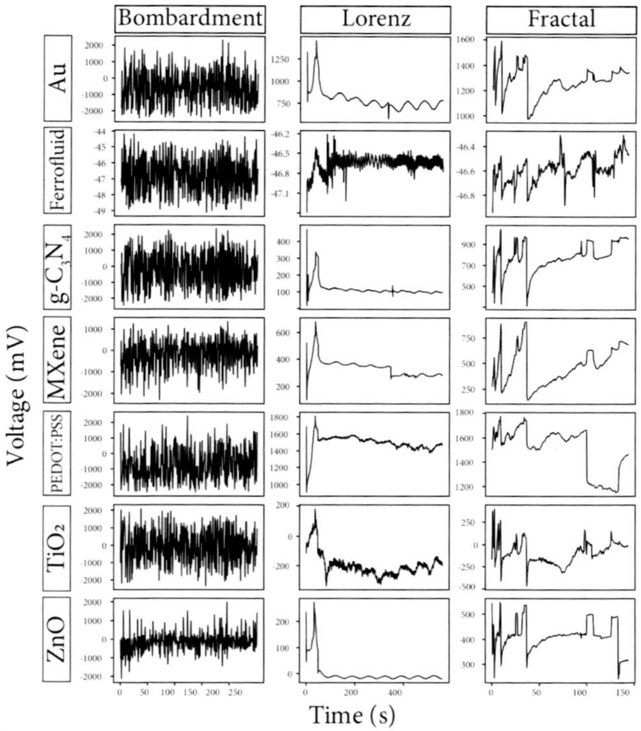

FIGURE 1
Measured electrical responses of the colloids for random, chaotic, and fractal signal inputs.

The colloid with the highest complexity across all inputs was TiO_2. For the fractal input, Au and PEDOT:PSS secured the second and third positions, while for the chaotic input, PEDOT:PSS and Au had the second and third positions, respectively. Finally, for the random binary string input, g–C_3N_4 and Au had the second and third positions. Fig. 1 shows the recorded electrical signals of the colloids for all inputs. Visually, the highly complex behavior of the TiO_2 colloid is evident, characterized by an unpredictable pattern.

Fig. 2(a) shows the measured electrical response for a chaotic signal input for the TiO_2 colloid. The inset within 2(a) zooms in on a region of the curve, highlighting the complex and unpredictable nature of the colloid's behavior when subjected to electrical signals. In addition, Fig. 2(b) shows the frequency-time wavelet spectrogram of the measured electrical response for the chaotic signal applied to the TiO_2 colloid. This spectrogram provides a dynamic representation of the response of the colloid over time, highlighting the distribution of frequencies and their variations. The predominantly dark plot, with limited areas of light shading, indicates a wide dispersion of

Colloid	Entropy	gzip Size	bzip2 Size	XZ Size	FI	FD
Fractal	3.48	6.13E-02	7.71E-02	8.80E-02	4.25E-01	1.31
Lorenz	4.50	1.00E-02	1.43E-02	2.16E-02	5.25E-01	1.00
Random	9.99E-01	2.38E-01	2.53E-01	3.56E-01	1.28E-01	1.98

TABLE 4
The calculated entropy, gzip size, bzip size, XZ size, maximal Lyapunov exponent, Fisher information, and fractal dimension of the input electrical signals

frequencies over time and a distinct lack of repetitive patterns in the data, highlighting the highly complex and dynamic nature of the response of the TiO_2 colloid and its dynamical behavior under electrical stimulation.

Tab. 4 shows the indexes calculated for the input electrical signals. As can be seen, the indexes are generally lower for the inputs than what has been observed for all colloid solutions, which indicates an increase in the complexity of the signal.

The increase in complexity was likely due to many processes occurring when an electric field is applied to the colloid. Ionizable and charged particles undergo translational movement and transient conduction due to the presence of an applied external field [5]. Dissociation of neutrals into charged species and recombination of ionized species take place in regions where the charge concentration deviates from equilibrium [27]. Charged species can also induce motion of uncharged neutral liquids, affecting charge transport and creating electrohydrodynamic conditions [3]. Particle transport can also arise from phenomena like dielectrophoresis [28], linear and nonlinear electrophoreses [19], and the Quincke rotational effect [29]. Simulating these dynamics is not an easy task [12,23,43,44] thus the measurements used here provide a fast way of selecting the most complex colloids without performing time-consuming simulations.

The observed increase in entropy along with file compression sizes suggests a potential use of colloidal solutions for reservoir computing. The correlation between entropy and classifier performance is well established in the machine learning community [4] and serves as a guiding principle for model optimization and performance evaluation [17, 18]. The observed trends in entropy and file compression sizes for the colloidal solutions for all inputs not only indicate possible good classification performance but also underscore the potential effectiveness of colloids in processing and extracting meaningful information from time series data. In addition, the data reveals a consistent trend in which the fractal dimension of the colloidal is generally lower than that of the input, suggesting the existence of memory within the colloidal solution [30]. This indicates the possibility of the usage of

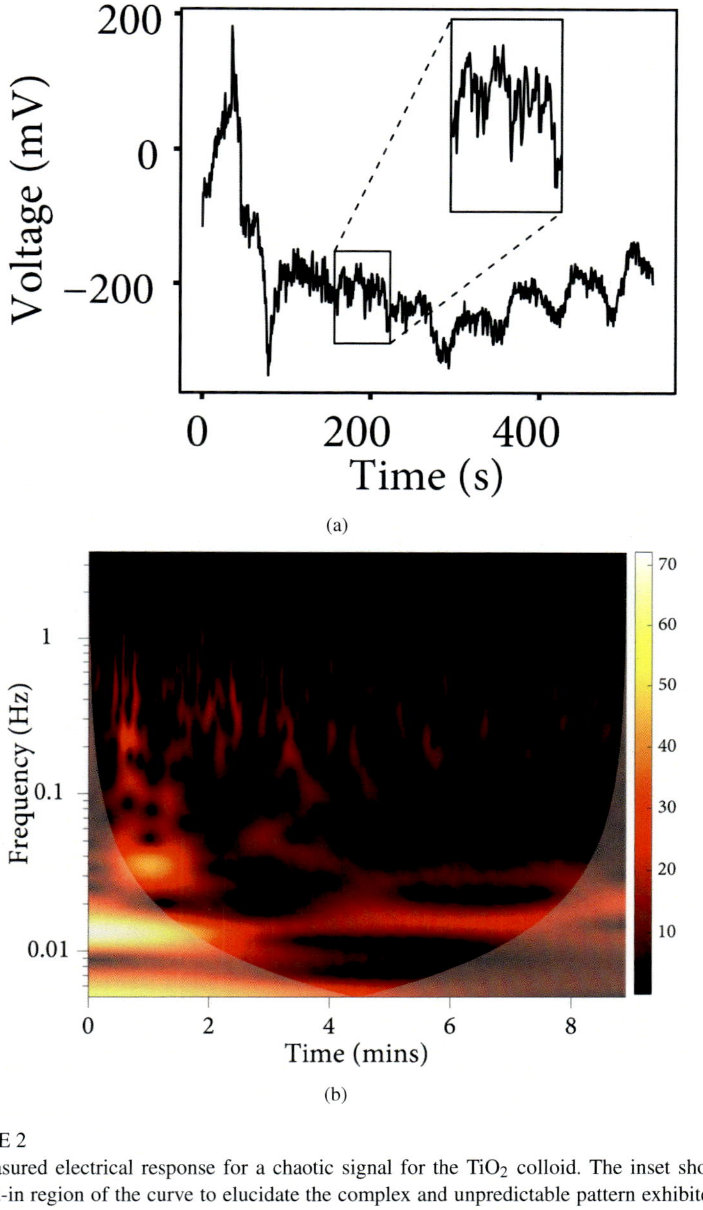

FIGURE 2
(a) Measured electrical response for a chaotic signal for the TiO_2 colloid. The inset shows a zoomed-in region of the curve to elucidate the complex and unpredictable pattern exhibited by the colloid when subjected to electrical signals. (b) Frequency-time wavelet spectrogram of the measured electrical response for a chaotic signal for the TiO_2 colloid. The predominantly dark plot with limited areas of light shading indicates a wide distribution of frequencies over time and a non-repetitive pattern in the data.

colloids as *in-materia* processing of time series data for high accuracy predictions [10, 14, 31, 34, 35].

5 BLACK-BOX NONLINEARITY

To measure the nonlinearity introduced by the colloid solution in the applied voltage, we define a measure of the nonlinearity. Given an input vector space V, an output vector space W, and a map $M : V \to W$, the "amount" of nonlinearity of the map in a region $\alpha \subseteq V$ is defined as

$$L_{\alpha}^{(M)} = \left\| M(v) - P(\hat{\theta}, v) \right\|_{\alpha} = \sqrt{\int_{\alpha} \left[M(v) - P(\hat{\theta}, v) \right]^2 \mathrm{d}v}, \qquad (4)$$

where $v \in V$ and $P(\hat{\theta}, v)$ is the hyperplane that best fits the data around the region α, and it is of the following form

$$P(\hat{\theta}, v) = \sum_{j} \hat{\theta^{j}} \cdot v^{j} + b, \qquad (5)$$

where $\hat{\theta}^{j} \in \hat{\theta}$ is the j-th estimated coefficient of the hyperplane, v^{j} is the j-th component of v, and b is an offset. In practice, we are sampling the region α around a discrete lattice, so it is safe to assume a finite number of vectors n. Then, the amount of nonlinearity can be defined as

$$L_{\alpha}^{(M)} = \sqrt{\sum_{i} \left[M(v_i) - P(\hat{\theta}, v_i) \right]^2}, \qquad (6)$$

where α is no longer a continuous region in the input space but a collection of input vectors $\{v_i\}, i = 1, \ldots, n$. Note that the summation in Eq. 6 may be interpreted as an approximation of the integral quantity defined in Eq. 4. Fundamentally, this nonlinearity measure shows that the output of the mapping M can be described as a linear combination of the input vectors plus a constant. Physically, this offset represents a possible potential present in the suspension when the measurement started.

Tab. 5 presents the estimated nonlinearity of the mapping of each colloidal system from the input stimulus to the measured output signal. Notably, the colloids that exhibited the highest complexity are also estimated to possess the highest amount of nonlinearity. In the context of reservoir computing, a crucial aspect of the reservoir is its capacity to perform a nonlinear mapping of the input data into a high-dimensional nonlinear state space. This

	$L_\alpha^{(M)}$		
Colloid	Fractal	Lorenz	Bombardment
Au	1.57	2.27	13.51
Ferrofluid	0.07	0.13	0.67
$g\text{–}C_3N_4$	2.32	2.19	15.61
MXene	3.15	4.20	9.24
PEDOT:PSS	3.54	2.56	15.43
TiO_2	7.42	8.14	14.52
ZnO	3.27	3.88	7.79

TABLE 5
Estimated nonlinearity of the output signal of the colloids, based on the input signals

nonlinear transformation of the input data enables the reservoir to capture and encode complex relationships and patterns present within the input data, allowing the subsequent linear readout layer to effectively extract and leverage these encoded features for tasks such as classification, regression, or prediction [40]. The high degree of nonlinearity of the colloids indicates their possibility for inclusion as nonlinear reservoirs in reservoir computing.

6 CONCLUSIONS

This work investigated the complexity and nonlinearity of various colloidal suspensions subjected to different electrical signal inputsfractal, chaotic, and random binary signals. Several complexity measures were applied to the recorded electrical output signals from the colloids, including entropy, file compression sizes, fractal dimension, and Fisher information. The most complex colloidal suspension across all input signals was found to be TiO_2, exhibiting high entropy, poor compressibility, and an unpredictable response pattern. The high complexity observed in the TiO_2 and other colloids suggests their potential usage for reservoir computing applications where transforming inputs into a high-dimensional nonlinear state space is desirable. To quantify the nonlinearity introduced by the colloids, a measure was defined based on the deviation of the output signal from the best-fit hyperplane of the input-output mapping. The results indicated that the colloids exhibiting higher complexity also possessed a higher degree of nonlinearity in their electrical responses. Overall, this study provides insights into the complex dynamics that can emerge in colloidal suspensions under electrical stimulation. The observed high complexity and nonlinearity highlight the potential of these unconventional computing materials for applications in neuromorphic computing, time series processing, and other domains that can exploit their rich dynamical properties. Further research is warranted to better understand the underlying mechanisms driving these complex phenomena and to explore their computational capabilities in greater depth.

REFERENCES

[1] Andrew Adamatzky. (2019). A brief history of liquid computers. *Philosophical Transactions of the Royal Society B*, 374(1774):20180372.

[2] Mohamed Alhabeb, Kathleen Maleski, Babak Anasori, Pavel Lelyukh, Leah Clark, Saleesha Sin, and Yury Gogotsi. (2017). Guidelines for synthesis and processing of two-dimensional titanium carbide (ti3c2t x mxene). *Chemistry of Materials*, 29(18):7633–7644.

[3] P. Atten. (October 1974). Electrohydrodynamic stability of dielectric liquids during transient regime of space-charge-limited injection. *The Physics of Fluids*, 17(10):1822–1827.

[4] Francisco J. Baldán and José M. Benítez. (March 2023). Complexity measures and features for times series classification. *Expert Systems with Applications*, 213:119227.

[5] Martin Z Bazant and Todd M Squires. (2010). Induced-charge electrokinetic phenomena. *Current Opinion in Colloid & Interface Science*, 15(3):203–213.

[6] Erik M. Bollt. (January 2017). Regularized forecasting of chaotic dynamical systems. *Chaos, Solitons & Fractals*, 94:8–15.

[7] R.L. Burden and J.D. Faires. (2010). *Numerical Analysis*. Cengage Learning.

[8] Alessandro Chiolerio. (2020). Liquid cybernetic systems: The fourth-order cybernetics. *Advanced Intelligent Systems*, 2(12):2000120.

[9] Marco Crepaldi, Charanraj Mohan, Erik Garofalo, Andrew Adamatzky, Konrad Szaciłowski, and Alessandro Chiolerio. (2023). Experimental demonstration of in-memory computing in a ferrofluid system. *Advanced Materials*, 35(23):2211406.

[10] Eugen Diaconescu. (2008). The use of NARX Neural Networks to predict Chaotic Time Series. *WSEAS Transactions on Computer Research*, 3(3).

[11] Yongchang Dong, Sai Sunil Kumar Mallineni, Kathleen Maleski, Herbert Behlow, Vadym N Mochalin, Apparao M Rao, Yury Gogotsi, and Ramakrishna Podila. (2018). Metallic mxenes: A new family of materials for flexible triboelectric nanogenerators. *Nano Energy*, 44:103–110.

[12] Suman Dutta and J. Chakrabarti. (2018). Transient dynamical responses of a charged binary colloid in an electric field. *Soft Matter*, 14(22):4477–4482.

[13] Raphael Fortulan, Noushin Raeisi Kheirabadi, Panagiotis Mougkogiannis, Alessandro Chiolerio, and Andrew Adamatzky. (2023). Reservoir computing with colloidal mixtures of zno and proteinoids. *arXiv preprint arXiv:2312.08130*.

[14] Indranil Ghosh, Manas K. Sanyal, and R. K. Jana. (August 2018). Fractal Inspection and Machine Learning-Based Predictive Modelling Framework for Financial Markets. *Arabian Journal for Science and Engineering*, 43(8):4273–4287.

[15] Kanit Hantanasirisakul, Mohamed Alhabeb, Alexey Lipatov, Kathleen Maleski, Babak Anasori, Pol Salles, Chanoknan Ieosakulrat, Pasit Pakawatpanurut, Alexander Sinitskii, Steven J May, *et al.* (2019). Effects of synthesis and processing on optoelectronic properties of titanium carbonitride mxene. *Chemistry of Materials*, 31(8):2941–2951.

[16] Chengmei Jiang, Cui Wu, Xunjia Li, Yao Yao, Lingyi Lan, Fengnian Zhao, Zunzhong Ye, Yibin Ying, and Jianfeng Ping. (2019). All-electrospun flexible triboelectric nanogenerator based on metallic mxene nanosheets. *Nano Energy*, 59:268–276.

[17] Yeliz Karaca and Dumitru Baleanu. (December 2020). A novel R/S fractal analysis and wavelet entropy characterization approach for robust forecasting based on self-similar time series modeling. *Fractals*, 28(08):2040032.

[18] Yeliz Karaca, Yu-Dong Zhang, and Khan Muhammad. (April 2020). Characterizing Complexity and Self-Similarity Based on Fractal and Entropy Analyses for Stock Market Forecast Modelling. *Expert Systems with Applications*, 144:113098.

[19] Aditya S. Khair. (June 2022). Nonlinear electrophoresis of colloidal particles. *Current Opinion in Colloid & Interface Science*, 59:101587.

[20] Noushin Raeisi Kheirabadi, Alessandro Chiolerio, Neil Phillips, and Andrew Adamatzky. (2023). Learning in colloids: Synapse-like zno+ dmso colloid. *Neurocomputing*, 557:126710.

[21] Noushin Raeisi Kheirabadi, Alessandro Chiolerio, Konrad Szaciłowski, and Andrew Adamatzky. (2023). Neuromorphic liquids, colloids, and gels: A review. *ChemPhysChem*, 24(1):e202200390.

[22] Noushin Raeisi Kheirabadi, Fathallah Karimzadeh, Mohammad Hossein Enayati, and Ehsan Naderi Kalali. (2023). Sustainable and photoresponse triboelectric nanogenerators based on 2d-gc3n4 and agricultural wastes. *Journal of Materials Science: Materials in Electronics*, 34(21):1571.

[23] K. S. Krishnamurthy, D. S. Shankar Rao, Madhu B. Kanakala, and Channabasaveshwar V. Yelamaggad. (April 2021). Electric response of topological dipoles in nematic colloids with twist-bend nematic droplets as the dispersed phase. *Physical Review E*, 103(4):042701.

[24] Mantas Lukoševičius and Herbert Jaeger. (2009). Reservoir computing approaches to recurrent neural network training. *Computer Science Review*, 3(3):127–149.

[25] Julian F Miller, Simon J Hickinbotham, and Martyn Amos. (2018). In materio computation using carbon nanotubes. In *Computational Matter*, pages 33–43. Springer.

[26] Julian Francis Miller. (2019). The alchemy of computation: designing with the unknown. *Natural Computing*, 18(3):515–526.

[27] V. Novotny. (January 1986). Contributions of particles to electrical conductivity of colloids. *Colloids and Surfaces*, 21:219–233.

[28] Oleg P. Pishnyak, Sergij V. Shiyanovskii, and Oleg D. Lavrentovich. (November 2011). Aggregation of colloidal particles in a non-equilibrium backflow induced by electrically-driven reorientation of the nematic liquid crystal. *Journal of Molecular Liquids*, 164(1):132–142.

[29] Gerardo E. Pradillo, Hamid Karani, and Petia M. Vlahovska. (August 2019). Quincke rotor dynamics in confinement: Rolling and hovering. *Soft Matter*, 15(32):6564–6570.

[30] Bo Qian and Khaled Rasheed. (2004). Hurst exponent and financial market predictability. In *IASTED conference on Financial Engineering and Applications*, pages 203–209. Proceedings of the IASTED International Conference Cambridge, MA.

[31] Bo Qian and Khaled Rasheed. (February 2007). Stock market prediction with multiple classifiers. *Applied Intelligence*, 26(1):25–33.

[32] Neda Raeisi-Kheirabadi and Alireza Nezamzadeh-Ejhieh. (2020). A z-scheme g-c3n4/ag3po4 nanocomposite: its photocatalytic activity and capability for water splitting. *International Journal of Hydrogen Energy*, 45(58):33381–33395.

[33] Noushin Raeisi Kheirabadi, Alessandro Chiolerio, and Andrew Adamatzky. (2024). Pavlovian reflex in colloids. *BioNanoScience*, pages 1–9.

[34] Sebastian Raubitzek and Thomas Neubauer. (May 2021). A fractal interpolation approach to improve neural network predictions for difficult time series data. *Expert Systems with Applications*, 169:114474.

[35] Sebastian Raubitzek and Thomas Neubauer. (November 2021). Taming the Chaos in Neural Network Time Series Predictions. *Entropy*, 23(11):1424.

[36] Nic Roberts, Noushin Raeisi Kheirabadi, Michail-Antisthenis Tsompanas, Alessandro Chiolerio, Marco Crepaldi, and Andrew Adamatzky. (2024). Logical circuits in colloids. *Royal Society Open Science*, 11(5):231939.

[37] C. E. Shannon. (July 1948). A mathematical theory of communication. *The Bell System Technical Journal*, 27(3):379–423.

[38] Marcelo Sobottka and Luiz P. L. de Oliveira. (2006). Periodicity and Predictability in Chaotic Systems. *The American Mathematical Monthly*, 113(5):415–424.

[39] Susan Stepney. (2019). Co-designing the computational model and the computing substrate. In *International Conference on Unconventional Computation and Natural Computation*, pages 5–14. Springer.

[40] Gouhei Tanaka, Toshiyuki Yamane, Jean Benoit Hroux, Ryosho Nakane, Naoki Kanazawa, Seiji Takeda, Hidetoshi Numata, Daiju Nakano, and Akira Hirose. (2019). Recent advances in physical reservoir computing: A review. *Neural Networks*, 115:100–123.

[41] Šime Ungar. (2007). The Koch Curve: A Geometric Proof. *The American Mathematical Monthly*, 114(1):61–66.

[42] Jiaxin Wu, Xin Jin, Shuo Mi, and Jinbo Tang. (June 2020). An effective method to compute the box-counting dimension based on the mathematical definition and intervals. *Results in Engineering*, 6:100106.

[43] Jiajia Zhou and Friederike Schmid. (November 2012). Dielectric response of nanoscopic spherical colloids in alternating electric fields: A dissipative particle dynamics simulation. *Journal of Physics: Condensed Matter*, 24(46):464112.

[44] Jiajia Zhou, Roman Schmitz, Burkhard Dünweg, and Friederike Schmid. (July 2013). Dynamic and dielectric response of charged colloids in electrolyte solutions to external electric fields. *The Journal of Chemical Physics*, 139(2):024901.

Liquid Spaceships for Advanced Autonomous Exploration

ALESSANDRO CHIOLERIO[1,2,*]

[1]*Bioinspired Soft Robotics, Istituto Italiano di Tecnologia, Via Morego 30, 16163 Genova, Italy*
[2]*Unconventional Computing Laboratory, University of the West of England, Bristol, BS16 1QY, UK*

Received: September 1, 2024. Accepted: September 28, 2024.

The dream of space exploration is approximately a century old, and still sparks admiration and awe among humanity, being supported by a growing economy that nurtures a fertile field where private and public cooperate towards the achievement of unexplored frontiers. Yet, manned exploration of distant worlds represents a far objective, and autonomous, robotic probes are developed to survive the harsh conditions found in space and to send back to Earth images and measurements of celestial bodies. Liquid spaceships here devised can offer substantial advantages, when compared to conventional architectures. A description of future hypothetical spaceships and missions is presented, based on the recent discoveries made in the field of liquid cybernetic systems.

Keywords: Liquid spaceship, liquid robotics, colloids, space exploration, deepspace, small satellites, gas giants, comets, asteroids

1 INTRODUCTION

We can state that space exploration started after the ideas of Konstantin Eduardovich Tsiolkovskii, whose pioneering work settled the concepts of the multi-stage rocket, of the life supporting systems for manned exploration, and of orbiting stations [44]. Several years after, during the cold war, the space program of the CCCP and of the USA begun, achieving incredible results and making exploration outside the Earth feasible. Due to the

* Corresponding author: E-mail: alessandro.chiolerio@iit.it

still huge costs involved in the launch and positioning in orbit, that scale with the mass of the specific vehicle, robotic exploration remains the most active field, as life support systems are bulky and heavy, and still suffer from drawbacks, if not show-stoppers, connected to equipment failure, radiation (or more generally space environment) exposure, and flight duration [12]. Yet, autonomous exploration systems undergo difficulties whenever highly charged particle fluxes, high temperatures, high pressures, uneven / anomalous gravito-electric fields are encountered [13, 21]. For example we might recall that autonomous probes undergo a freezing to protect their electronics from the higher charged particle densities whenever a strong magnetic field is crossed: this happens across the van Allen belts [33], across the Jovian belts [38] and also in the proximity of the Sun [35]. High temperatures like those in the proximity of the Sun or in the depth of the gas giants atmospheres [8] can prevent even military grade electron devices from operating, so that new architectures must be found [31]. High pressures can be lethal for an autonomous exploration system, we might recall the collapse of the Probe mission after diving into Jupiter's clouds [47]. Uneven gravitational fields make it difficult to remotely operate an orbiting probe [22], and constitute an extreme difficulty also for landing probes [7].

Nevertheless, current missions are targeting exotic bodies for several fundamental reasons: mining rare raw materials, understanding eventual harms to our planet, or simply providing scientific observations in search for extraterrestrial life [51]. This makes of paramount importance exploring new paradigms to provide fresh exploration means. Even soft robotics has become attractive for robotics in some specific domains: a soft pellicle, allowing mechanical deformation and stretching, can easily follow the morphing of a structure that adapts to the environment and exchanges forces to provide locomotion, even in solid or liquid environments [32]. Space exploration with soft robots has been considered as well [57]. Liquid robotics does not represent only a subset of soft robotics (functional liquids confined by a soft membrane) but also an extension of its borders, being a field where all the functionalities of an autonomous system can be sustained by a same liquid state material. To predict the governing laws that describe the interaction between an agent (the spaceship in our case) and the surrounding environment we recall the third order cybernetic paradigm [11] while considering critical situations when the agent redefines itself in a dynamical dialogue with the universe we recall the fourth order [15] or the metacybernetic description [55]. The chapter is organized in the following sections: section 2 lists the subsystems of a completely autonomous liquid spaceship and describes their functioning; section 3 discusses (Earth) orbiting systems, the exploration of small bodies, of gas giants, of distant bodies in the deepspace; section 4 closes the chapter.

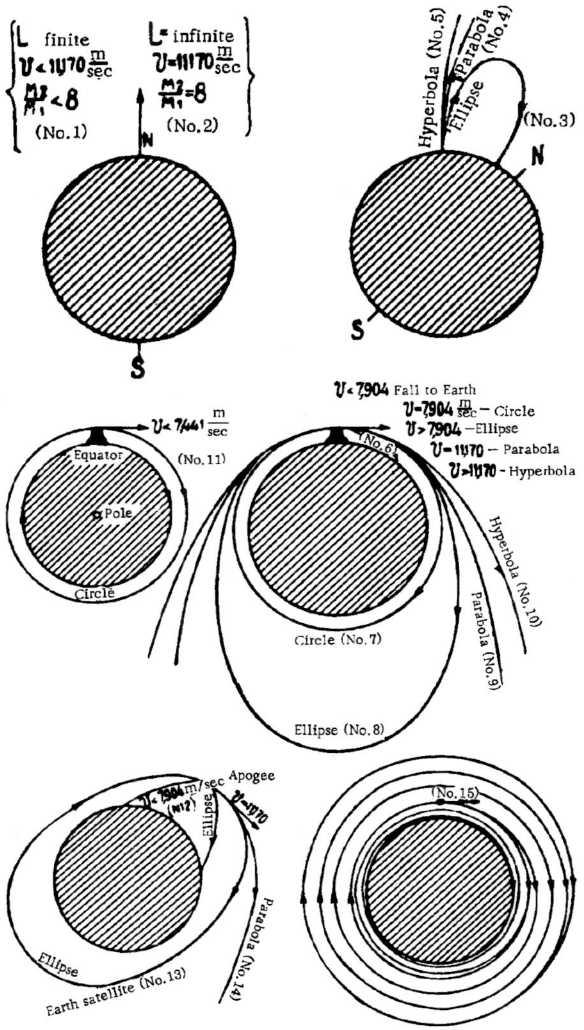

FIGURE 1
Different flight trajectories drawn by K.E. Tsiolkovskii ("Vestnik Vozdukhoplavaniya", No. 2 and 3, 1912) [44].

2 THE STRUCTURE OF A LIQUID SPACESHIP

A typical scenario we are considering is that of a conventional launch that can deliver a liquid state spaceship into the transfer orbit, directed towards its objectives. The components will be subjected to a targeted design on the basis of the physical parameters found in the target proximity, so that the functional subsystems can activate and operate when necessary. For example in case the target environment features a specific temperature range, the solvent choice must be such that freezing and boiling points are duly avoided. A skin that prevents liquid loss by evaporation, ablation, or accidental contacts can be designed using a proper choice of materials that soften at the working temperature, including glass, metals and related alloys, polymers and polymeric matrix composites. In a particular embodiment of a liquid spaceship, we might imagine to use surfactants to create a tight shell, whose intermolecular distances do not allow the diffusion and evaporation of the solvent molecules. However we might expect that physical parameters of the external environment will influence the surfactant features. For example temperature affects the curvature of surfactants and therfefore their aggregation liquid crystal structure. Electrical charge also produces swelling and increase in the diameter of the solvent channels [50]. Even pH variations influence molecular curvature [40]. Conventional frames and electronics are not excluded from the spaceship structure, therefore hybrid vehicles might offer the best performances. As an example, a conventional mother-ship could host the liquid daughter-ship in a proper reservoir, and release all (or a portion) of its content when needed, providing energy, data relay, telemetry, and so on, from its orbiting trajectory. For a detailed analysis of materials and components of a liquid spaceship, please refer to [14]. The spaceship features the following fundamental subsystems.

2.1 The energy subsystem

Energy generation and storage are of paramount importance to power a spaceship. Energy could be beamed by a ground station on Earth if the spaceship is travelling in line-of-sight or close to the surface, for example using high frequency electromagnetic waves or laser beams. A liquid opaque to the specific field would then convert the oscillations into displacement, changes in its electrical charge or magnetic momentum, that could be converted by engineering system. In another scenario the energy must be harvested from a local gradient in the target environment: a thermal, electric, magnetic, gravitational, chemical, kinetic one [17]. For example a thermal gradient represents a quite ubiquitous situation, where light and shadow cast a terminator on the astronomical object under exploration / on the spaceship. By exploiting photothermal conversion, light can be converted into a thermal gradient [6], and the thermal gradient into energy that can be used by the spaceship, through at least one among the possible physical effects known [5]. Ultimately, self-generation is also a possible route for producing energy, for example consuming a chemical reaction (a redox) like in a solid state battery, or converting a radioactive decay into a current or again a thermal gradient, similarly to Radiation Thermoelectric Generators (RTGs) used in current deep space

FIGURE 2
Liquid energy harvester hallucination provided by AI (Dalle 3).

missions. Converting energy by itself does not suffice to the requirements of space exploration, as an energy management layer is also necessary to enhance conversion efficiency, particularly in case of discontinuous generation, of voltage scaling or pumping to adapt to the processing requirements, of storage in case of overproduction [56]. Such management tasks must be implemented by hybrid, silicon-based systems as of today. Energy distribution, necessary in case the spaceship has a considerable size, can be implemented either by passive systems, such as thermally or movement- induced motion of energetic fluids inside soft pipelines (see Figure 2), or by active pumping.

2.2 The mobility subsystem

Moving a spaceship with respect to a reference frame is fundamental. During navigation, fixed stars are used and pointed by the star trackers for the correct tracking of the vessel [3]. An autonomous system usually performs gravitational assists, by exchanging momentum with a planetary system, to save fuel and gain speed, allowing the exploration of more external orbits, for example [34]. It might happen that the encounter for observation is performed at a relatively high speed, so that carefully planned measurements can be done, with no maneuvers: the spaceship is a bullet shot close to the objective, like the Deep Space New Horizons sent to Pluto [24]. Specific objectives whose size, inertia, and trajectory cannot be predicted with details, require an approach, an observation and a trajectory correction with decision autonomy [37]. A liquid spaceship still requires conventional propulsion means (cryogenic, chemical, plasma, and so on), but makes possible two aspects: 1) reorientation and 2) morphing. Reorientation can be done by controlling the toroidal movement of a portion of liquid in a spherical shell: having a colloid composed by polarizable particles (either susceptible to the magnetic or the electric field), and means to locally generate the controlling field, constitutes a liquid reaction wheel [10]. Other easier implementations encompass pumping systems, valves, and pipelines. Pumping, besides conventional propellers, can be obtained by inducing a motion in active colloids containing microswimmers: particles designed with the specific task of manifesting phototaxis (movement induced by light), chemotaxis (here the motion is triggered by a chemical reaction occurring asymmetrically on the surface of bi-anisometric particles, the so called Janus particles), thermotaxis (movement induced by a thermal gradient), and so on [39]. Morphing is possible using similar means of inducing movements in a liquid, more specifically such movements are conceived to deplete some sections of the spaceship and to fill some other sections, therefore producing a variable stiffness in the segments [25]. This is a biomimetic approach that can be used to produce worm-like propulsion capabilities, and would be particularly beneficial in those environments where the local fields are faint and normal forces cannot be exploited for locomotion. Furthermore morphing allows re-orientation of portion(s) of the autonomous system, that becomes capable of pointing the sensing section towards caves, craters, cavities which are not in line-of-sight when the system is normally oriented. Conventional propulsion is also possible, in a particular scenario where the functional matter that realizes the spaceship is consumed as propellant, expelled to obtain a physical thrust. This will leave a molecular footprint of the trajectory of the spaceship, traceable as the Ariadne's thread. An example is given by ferrofluids: in the presence of a magnetic field, a ferrofluid surface deforms and creates a spiky structure, known as Rosenweig instability; ionic ferrofluids further respond to a proper electric field initiating an electrospray emission that has been developed for micropropulsion [46].

2.3 The sensing subsystem

Measuring physical quantities can be performed with two different frameworks: 1) volume sensing and 2) local sensing. The first approach leverages on changes in the physical properties of the liquid body of a spaceship, when it is submitted to a scalar or vectorial field whose gradient has a spatial component far bigger than the spaceship size. Examples encompass the atmospheric pressure when diving into the depths of a gas giant, or the magnetic field generated by a star, or the radiation field incoming from a distant object. We just provide some practical examples of possible implementations for each of this scenarios. The hydrostatic pressure pushes against the skin of the liquid spaceship: by incorporating hollow structures in the skin, and filling them with electrically conductive liquids such as ionic liquids [54] or liquid metal alloys [23], it is possible to measure a pressure-induced variation in the capacitance [59] or in the inductance of microdevices [19], giving to the spaceship also a spatial resolution that can provide a measure of the effects of strong winds, for example. Liquid antennas can also provide an interesting system for realizing distributed pressure sensors [58]. Sensitivity to the environmental magnetic field is granted by integrating (super-para)magnetic nanoparticles into the liquid composition, such that the magnetic field strength can change the density of the particles and the strength of their dipolar interaction, ultimately changing the particle dynamics. For example integrating a microsystem able to perform Dynamic Light Scattering (DLS) analyises on the magnetic particles [27] provides information that can be decoupled to reconstruct the external magnetic field. The radiation flux to which a spaceship is exposed includes electromagnetic radiation, electrons and heavier particles such as protons and ionised nuclei. Therefore electrical charge might travel across the spaceship skin, depending on the kinetic energy of the particles, and be accumulated within the liquid body. Furthermore, particularly when the incoming radiation features a high energy in the MeV range, secondary radiation might be emitted by the liquid itself, initiating a decay process. Most probably, all of the electromagnetic properties of colloids can be affected by such decay. During experiments performed at CNAO, a commercial ferrofluid (EMG601p from Ferrotech, U.S.A.) was exposed to packets of protons at an energy of 227 MeV for a total amount of 9.1×10^{11} protons, in the facilities of the National Center for Oncological Hadrontherapy in Pavia, Italy. Measuring the voltage spontaneous fluctuations within the liquid sample before, during the exposure, and after one day, produces very different patterns in the dynamics of the noise. The Fast Fourier Transform (FFT) analysis provides an immediate tool for appreciating such differences, as shown in Figure 3.

2.4 The data processing and storage subsystem

Despite the amorphous nature of liquids and colloidal suspensions, having understood the advantages offered by the physical aggregation state in terms of shape adaptability, mechanical stress absorption, electrostatic discharge dispersion and ionizing radiation resilience, in recent years both the processing and the storage capabilities of such materials have been explored. Providing that the solvent is liquid in the temperature range required by the operations, liquid processors, memories and logics are superior in terms of energy efficiency and therefore could be fundamental in providing continuous, resilient operation in deepspace explorations, where little or no energy

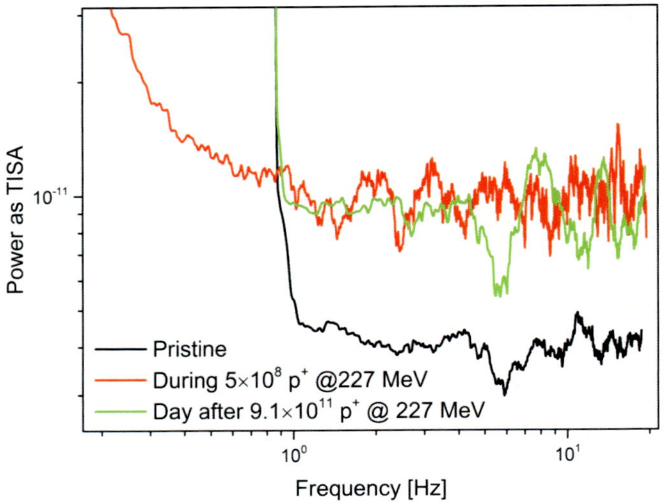

FIGURE 3
Fast Fourier Transform analysis of the different patterns in the spontaneous fluctuations of the voltage measured across a ferrofluid sample exposed to an intense radiation dose (as indicated in the inset). The fluctuations power is measured as Time-Integral Square Amplitude in logarithmic scale.

can be harvested. Energy efficiency stems out from the intrinsic complexity offered by the degrees of freedom of the liquid molecular structure, where base constituents show also quantum-related properties such as magnetization. Dielectric polarizability, volume extension versus molecular folding, the magnetization itself, can be influenced by local fields. The processor and memory should be located anyhow in a protected volume of the spaceship, even if their capabilities can be easily distributed over the entire spaceship volume, exactly as it happens for the information content of a holographic memory. Electrically programmable in-memory computing can be performed with liquids, exploiting the re-configuration that electrical stimuli provide to the meta-stable steric configuration within a ferrofluid [20]. A memristive response is found, encompassing both short and long-term information storage capabilities as well as plasticity. Physical reservoir computing through training a readout layer has also been accomplished with the same ferrofluid. Furthermore the demonstration of learning pathways has been achieved leveraging on the dynamical aspects of electronic or ionic conductive nanoparticles (zinc oxide) dispersed in a dielectric solvent (dimethyl sulfoxide) [30]. Nonetheless, such learning pathways are transient, they deteriorate over time. Biological paradigm such as the Pavlov conditional reflex have also been observed in colloids [29], whose plasticity is normally based

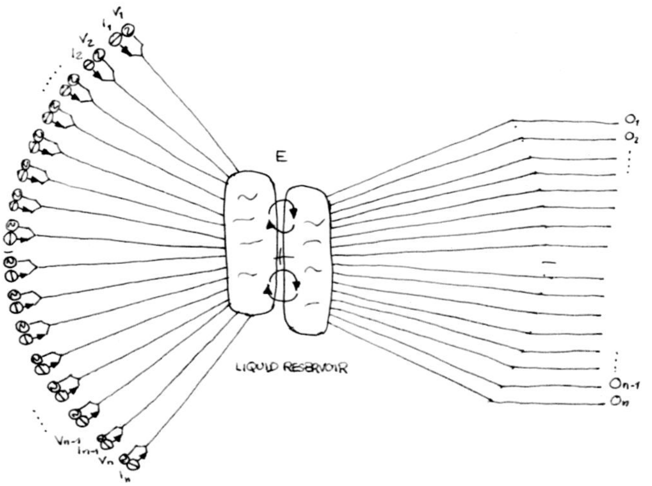

FIGURE 4
Innovative computational scheme where a liquid reservoir is used both as high-dimensional projector and as a tuneable weigth matrix. On the left portion the n inputs composed by signals (the proper input V_n) and current generators (that force the weights I_n). On the right portion the outputs (voltage readings O_n). The liquid reservoir is in this case schematically represented in its entangled "E" state, where two physically separated volumes share a phase correlation through which they can behave as one single entity. Fountain pen-drawn on light paper.

on chemo/physical mechanisms that induce metastable states [16]. Emulating complex mechanisms such as those found in biological neural networks therefore appears feasible. Even logical circuits, the Boolean gates, have been discovered by submitting liquid systems to electrical impulses [43], using two, four, and eight inputs. All such characterization setups require metallic electrodes in contact with the liquid reservoir, plus a conventional, silicon-based device where instructions can run, for example including an Artificial Neural Network. Nonetheless all-liquid schemes could still be feasible, where the reservoir provides not only a projection into a vast phase space of the input instructions, but also the plasticity to provide a matrix of tuneable weights, as proposed in the sketch of Figure 4. There, the input is composed by a parallel of signal generators (the proper input) and voltage controlled current generators (that force weights). Outputs are extracted by measuring the voltage fluctuations of the reservoir.

2.5 The communication subsystem

Communication is fundamental for data relay and for operation, when the distance of the target point from Earth is compatible with the decisions to be taken during exploration. Another form of communication takes place between components of the observer, namely an orbiter and a lander or multiple orbiters, if it is the case, or among different agents of a swarm robotic system. The orbiter / lander configuration has been successfully exploited in planetary exploration several times [4]. Multiple orbiters have been exploited for example to map vectorial field features simultaneously from spatially resolved locations [45]. Liquid antennas offer the advantage of being easily reshaped, in order to point and focus toward their target, by the application of a magnetic field for example. They can be folded during launch to help in saving volumes and morphed into position when necessary. Current research has proposed the use of liquid mirrors both on Earth applications [2, 9] and also for orbiting telescopes [1]. Possible colloids include ferrofluids and plasmonic nanoparticles [42]. Also liquid antennas have been widely explored [26]. Recent findings show that a particular property of liquids, namely a phase correlation between different volumes of ferrofluids, kept separated and properly insulated from the electromagnetic field, provides a potentially exploitable effect [18]. What happens is that the coherent condensate structure of the system ground state allows quantum dynamics at the microscale to emerge and provoke a macroscopic behavior. As far as we know the effect is found in liquids, therefore in a specific temperature range, where it is not overcome by quantum decoherence. This means that it is possible to influence the electrodynamic status of one spaceship from another liquid agent, previously twinned, and viceversa. This property has features in common with quantum communication and allows copying data without the need for line-of-sight alignments, opening new options for liquid swarm autonomous explorers and fusion sensing techniques [28]. Going back to Figure 4 we can now understand why the reservoir has been depicted in the entangled state: physically detached reservoir sub-components can be used to perform reservoir computing, regardless of their position in space-time. The results of the computation is therefore dependent on the different environments where sub-components are located (quantum sensing).

3 EXPLORATION SCENARIOS

Orbiting around Earth is an application of paramount importance, particularly from the point of view of the economics that are fuelling space technologies for military and civilian applications. A liquid satellite offers several advantages, for the small sat field (nano, pico, cube [41]): they could be ejected using a single mother-reservoir that delivers them in the due location. They are resilient to space debris, since an object moving at a high relative speed could hit, trespass, and proceed beyond the liquid satellite without causing a dispersion of its constituents. Surface tension would rearrange the body giving it a spherical symmetry, and surfactant molecules would float until the surface to reconstruct their protecting membrane.

Small bodies such as a comet, an asteroid, a minor natural satellite, feature a faint gravitational field whose effects on the probe are comparable to those induced by the local electric / magnetic fields [48]. This environment encompasses extremely low temperatures, sudden jets of plasma-like gases or charged dust, uneven surfaces with rocks / dust pile areas, crater or cave structures to be explored. For example cryovolcanic vents on Pluto, ice fissures on the Jupiter's moons Enceladus and Europa, on Saturn moon Titan, and on Uranus moon Triton, holes at the bottoms of impact craters on Mercury, several thousands of cave entrances identified on the Moon and on Mars [53]. A liquid exploration vehicle, endowed with a proper protection membrane, will control its shape, relative positioning and thrust to properly navigate in the faint gravitational field provided by the small body under analysis.

At odds, a gas giant with a diameter bigger than the Earth's will feature an opaque atmosphere, below which a potential liquid and / or solid surface might exist. This environment encompasses huge pressures and temperatures, strong winds, electrical storms, corrosive atmospheres, surfaces to be explored [49], and is among the least known places, holding a huge potential for discoveries. As depicted in Figure 5, a liquid planetary explorer will travel by conventional means until the insertion orbit of the gas giant, to be deployed and start approaching the cloud layer. Their systems will be activated during descent, providing coarse readings of electromagnetic emissions and other physical parameters. By calculating the proper synthetic cytoplasm density it will be possible to adapt the buoyancy and reduce the entry speed.

The discovery of exoplanets has shown that solar systems in our galaxy, the Milky Way, are much more varied than we once thought. Some, like those with gas giants orbiting close to their stars, are very different from our solar system. This suggests that the way planets form, move, and evolve is not fully understood and may differ across the galaxy. By studying the minerals, isotopes, and molecules in these systems, we can learn more about the conditions under which they formed and how they evolved [36]. Beyond telescopic observations we would need to know the ratios of isotopes (e.g., deuterium/hydrogen) and the amounts of the elements, which can only be gathered through direct measurements. Even traveling to the closest star system, Alpha Centauri (4.2×10^{13} km), could take hundreds or thousands of years. In a typical deepspace mission a liquid spaceship will be directed either to an object in the periphery of the Solar system, for example orbiting in the Oort cloud [52], or to a close star. This environment encompasses extremely low temperatures with negligible levels of light radiation coming from the Sun, potentially high charged particle fluxes and other unknown distribution of force fields along the probe path. Let us imagine that the liquid spaceship is

FIGURE 5
Liquid cloud explorer hallucination provided by AI (Nightcafé).

at macroscopic level completely homogeneous and has a flight configuration resembling a droplet, eventually preserved in a conventional shell that integrates communication antennas, standard electronics, and other fundamental subsystems. The acceleration to transfer speeds will be achieved either by liquid propulsion or by other more conventional means, until cruise speed is reached. Normally this is considered to be a fraction of light speed, such that a tremendous amount of momentum has to be released to slow down in the proximity of the target. While a conventional system has little options, rather than decelerating with the same propulsion apparatus, a liquid spaceship has the option of expanding its cross-section orthogonal to the speed vector, to intercept celestial bodies or to be captured by their gravitational fields. The expansion could be triggered by heat (thermotaxis) or radiation (phototaxis) in the destination proximity and could also be a slow process, taking years to complete. The random capture process will fragment the spaceship and disperse its material, so that by exploiting phase correlation, the system working as a whole, information could be retrieved and sent back to Earth.

4 CONCLUSIONS

We have seen how the exploration of outer space still poses several challenges, that might be properly addressed (and solved) by designing a liquid state autonomous exploration vehicle. Besides engineering features that will allow liquid explorers to survive to intense radiation doses, strong magnetic fields, and less predictable environments, several unique features will be enabled by liquid space exploration, including scission, fusion, shape morphing and phase-correlated swarm exploration. The unique features of liquids and the way they respond to external stimuli can be exploited both in the proximity applications (Earth observation) and in the deepspace frontier.

ACKNOWLEDGEMENTS

This project has received funding from the European Innovation Council And SMEs Executive Agency (EISMEA) under grant agreement No. 964388. We would like to thank Dr. Marco B. Quadrelli for inspiring this study, Dr. Maurice Yolles, and Dr. Satyajit Ravindra for fruitful discussions, Dr. Marco Pullia for hosting the measurements under proton beam at CNAO, the use of the CNAO experimental facility built in collaboration with INFN, and Prof. Andrew Adamatzky for hosting the Gagarin's Readings.

REFERENCES

[1] What is the fluidic telescope? https://www.nasa.gov/science-research/astrophysics/what-is-the-fluidic-telescope/. Accessed: 2024-09-22.

[2] (2006). A liquid mirror. *Science*, 312(5777):1107–1107.

[3] (2018). Chapter 6 - advanced space optical attitude sensor. In Zheng You, editor, *Space Microsystems and Micro/nano Satellites*, Micro and Nano Technologies, pages 167–231. Butterworth-Heinemann.

[4] Andrew Ball, James Garry, Ralph Lorenz, and Viktor Kerzhanovich. (2007). *Planetary Landers and Entry Probes*. Cambridge University Press.

[5] M. Bevione, L. Cecchini, E. Garofalo, S. A. Suslov, and A. Chiolerio. (2023). *Colloidal Technologies for Heat Energy Recovery*, pages 49–104. Springer International Publishing, Cham.

[6] Matteo Bevione, Alessandro Chiolerio, and Giulia Tagliabue. (2023). Plasmonic nanofluids: Enhancing photothermal gradients toward liquid robots. *ACS Applied Materials & Interfaces*, 15(43):50106–50115. PMID: 37853519.

[7] Hermann Boehnhardt, Jean-Pierre Bibring, Istvan Apathy, Hans Ulrich Auster, Amalia Ercoli Finzi, Fred Goesmann, Göstar Klingelhöfer, Martin Knapmeyer, Wlodek Kofman, Harald Krüger, Stefano Mottola, Walter Schmidt, Klaus Seidensticker, Tilman Spohn, and Ian Wright. (2017). The philae lander mission and science overview. *Philosophical Transactions of the Royal Society A: Mathematical, Physical and Engineering Sciences*, 375(2097):20160248.

[8] S. J. Bolton, A. Adriani, V. Adumitroaie, M. Allison, J. Anderson, S. Atreya, J. Bloxham, S. Brown, J. E. P. Connerney, E. DeJong, W. Folkner, D. Gautier, D. Grassi, S. Gulkis, T. Guillot, C. Hansen, W. B. Hubbard, L. Iess, A. Ingersoll, M. Janssen, J. Jorgensen,

Y. Kaspi, S. M. Levin, C. Li, J. Lunine, Y. Miguel, A. Mura, G. Orton, T. Owen, M. Ravine, E. Smith, P. Steffes, E. Stone, D. Stevenson, R. Thorne, J. Waite, D. Durante, R. W. Ebert, T. K. Greathouse, V. Hue, M. Parisi, J. R. Szalay, and R. Wilson. (2017). Jupiter's interior and deep atmosphere: The initial pole-to-pole passes with the juno spacecraft. *Science*, 356(6340):821–825.

[9] Ermanno F. Borra, Grégoire Tremblay, Yannick Huot, and Jonny Gauvin. (1997). Gallium liquid mirrors: Basic technology, optical-shop tests and observations. *Publications of the Astronomical Society of the Pacific*, 109(733):319–325.

[10] Miguel Castroviejo-Fernandez and Ilya Kolmanovsky. (2023). Constrained reaction wheel desaturation and attitude control of spacecraft with four reaction wheels. *IFAC-PapersOnLine*, 56(2):6005–6011. 22nd IFAC World Congress.

[11] Evgeny Chepin. (2021). Robotics: From first-order cybernetics to third-order cybernetics. *Procedia Computer Science*, 190:130–136. 2020 Annual International Conference on Brain-Inspired Cognitive Architectures for Artificial Intelligence: Eleventh Annual Meeting of the BICA Society.

[12] Steve Chien and Kiri L. Wagstaff. (2017). Robotic space exploration agents. *Science Robotics*, 2(7):eaan4831.

[13] Steve A. Chien, Gianfranco Visentin, and Connor Basich. (2024). Exploring beyond earth using space robotics. *Science Robotics*, 9(91):eadi6424.

[14] A. Chiolerio and Marco B. Quadrelli. (2017). Smart fluid systems: The advent of autonomous liquid robotics. *Advanced Science*, 4(7):1700036.

[15] Alessandro Chiolerio. (2020). Liquid cybernetic systems: The fourth-order cybernetics. *Advanced Intelligent Systems*, 2(12):2000120.

[16] Alessandro Chiolerio, Erik Garofalo, Neil Phillips, Ermelinda Falletta, Rodrigo de Oliveira, and Andrew Adamatzky. (2024). Learning in colloidal polyaniline nanorods. *Results in Physics*, 58(107501).

[17] Alessandro Chiolerio and Marco B. Quadrelli. (2019). Colloidal energetic systems. *Energy Technology*, 7(5):1800580.

[18] Alessandro Chiolerio, Giuseppe Vitiello, Mohammad Mahdi Dehshibi, Marco Crepaldi, Diego Torraza, and Andrew Adamatzky, (07 2024). Room temperature multi-particle entanglement in ferrofluids.

[19] Jungkeun Choi, Seokbeom Kim, Jungchul Lee, and Bumkyoo Choi. (2015). Improved capacitive pressure sensors based on liquid alloy and silicone elastomer. *IEEE Sensors Journal*, 15(8):4180–4181.

[20] Marco Crepaldi, Charanraj Mohan, Erik Garofalo, Andrew Adamatzky, Konrad Szaciłowski, and Alessandro Chiolerio. (2023). Experimental demonstration of in-memory computing in a ferrofluid system. *Advanced Materials*, 35(23):2211406.

[21] Yang Gao and Steve Chien. (2017). Review on space robotics: Toward top-level science through space exploration. *Science Robotics*, 2(7):eaan5074.

[22] Romain Garmier, Jean-Pierre Barriot, Alexander S. Konopliv, and Donald K. Yeomans. (2002). Modeling of the eros gravity field as an ellipsoidal harmonic expansion from the near doppler tracking data. *Geophysical Research Letters*, 29(8):72–1–72–3.

[23] Osman Gul, Kyuyoung Kim, Jimin Gu, Jungrak Choi, Dionisio Del Orbe Henriquez, Junseong Ahn, and Inkyu Park. (2021). Sensitivity-controllable liquid-metal-based pressure sensor for wearable applications. *ACS Applied Electronic Materials*, 3(9):4027–4036.

[24] Y. Guo and R.W. Farquhar. (2008). New horizons mission design. *Space Sci Rev*, 140:49–74.

[25] Zicun Hong, Haoyu Huang, and Yong Zhong. (2023). A novel variable stiffness underwater biomimetic propulsion mechanism. In *2023 29th International Conference on Mechatronics and Machine Vision in Practice (M2VIP)*, pages 1–6.

[26] Yi Huang, Lei Xing, Chaoyun Song, Stephen Wang, and Fatma Elhouni. (2021). Liquid antennas: Past, present and future. *IEEE Open Journal of Antennas and Propagation*, 2:473–487.

[27] Jingyu Jin, Dongxing Song, Jiafeng Geng, and Dengwei Jing. (2018). Time-dependent scattering of incident light of various wavelengths in ferrofluids under external magnetic field. *Journal of Magnetism and Magnetic Materials*, 447:124–133.

[28] Bahador Khaleghi, Alaa Khamis, Fakhreddine O. Karray, and Saiedeh N. Razavi. (2013). Multisensor data fusion: A review of the state-of-the-art. *Information Fusion*, 14(1):28–44.

[29] Noushin Raeisi Kheirabadi, Alessandro Chiolerio, and Andrew Adamatzky. (2024). Pavlovian reflex in colloids. *BioNanoScience*, Article in press.

[30] Noushin Raeisi Kheirabadi, Alessandro Chiolerio, Neil Phillips, and Andrew Adamatzky. (2023). Learning in colloids: Synapse-like zno+ dmso colloid. *Neurocomputing*, 557:126710.

[31] H.K. Kim. (2019). Vacuum transistors for space travel. *Nat Electron*, 2:374–375.

[32] Tuck-Whye Wong Benjamin Shih Chunyu Guo Luwen Wang Jiaqi Liu Tao Wang Xiaobo Liu Jiayao Yan Baosheng Wu Fajun Yu Yunsai Chen Yiming Liang Yaoting Xue Chengjun Wang Shunping He Li Wen Michael T. Tolley A-Man Zhang Cecilia Laschi & Tiefeng Li Li, G. (2023). Bioinspired soft robots for deep-sea exploration. *Nat Commun*, 14(7097).

[33] W. Li and M.K. Hudson. (2019). Earth's van allen radiation belts: From discovery to the van allen probes era. *Journal of Geophysical Research: Space Physics*, 124(11):8319–8351.

[34] Xin Ma, Jiancheng Fang, and Xiaolin Ning. (2013). An overview of the autonomous navigation for a gravity-assist interplanetary spacecraft. *Progress in Aerospace Sciences*, 63:56–66.

[35] D.J. McComas, E.R. Christian, and C.M.S. (2019). Probing the energetic particle environment near the sun. *Nature*, 576:223–227.

[36] Kimberly Moore, Samuel Courville, Sierra Ferguson, Ashley Schoenfeld, Kristie Llera, Rachana Agrawal, Daniel Brack, Peter Buhler, Kyle Connour, Ellen Czaplinski, Michael DeLuca, Ariel Deutsch, Noah Hammond, Donald Kuettel, Angela Marusiak, Stefano Nerozzi, Jeffrey Stuart, Jesse Tarnas, Alexander Thelen, Julie Castillo-Rogez, William Smythe, Damon Landau, Karl Mitchell, and Charles Budney. (2021). Bridge to the stars: A mission concept to an interstellar object. *Planetary and Space Science*, 197:105137.

[37] I.A. Nesnas, L.M. Fesq, and R.A. Volpe. (2021). Autonomy for space robots: Past, present, and future. *Curr Robot Rep*, 2:251–263.

[38] Q. Nénon, A. Sicard, P. Kollmann, H. B. Garrett, S. P. A. Sauer, and C. Paranicas. (2018). A physical model of the proton radiation belts of jupiter inside europa's orbit. *Journal of Geophysical Research: Space Physics*, 123(5):3512–3532.

[39] Audrey Nsamela, Aidee Itandehui Garcia Zintzun, Thomas D. Montenegro-Johnson, and Juliane Simmchen. (2023). Colloidal active matter mimics the behavior of biological microorganisms—an overview. *Small*, 19(13):2202685.

[40] Julian Philipp, Aleksandra Dabkowska, Anita Reiser, Kilian Frank, Rafał Krzysztoń, Christiane Brummer, Bert Nickel, Clement E. Blanchet, Akhil Sudarsan, Mohd Ibrahim, Svante Johansson, Pia Skantze, Urban Skantze, Sofia Östman, Marie Johansson, Neil

Henderson, Kjetil Elvevold, Bård Smedsrød, Nadine Schwierz, Lennart Lindfors, and Joachim O. Rädler. (2023). ph-dependent structural transitions in cationic ionizable lipid mesophases are critical for lipid nanoparticle function. *Proceedings of the National Academy of Sciences*, 120(50):e2310491120.

[41] Dušan Radosavljević, Lazar Jeftić, L. V. Muralikrishna Reddy, K. Gopalakrishnan, and S. Mohankumar. (2021). Era of small satellites: Pico, nano and micro-satellites (pnm sat)—an over view of frugal way to access low earth orbit. In Devendra Kumar Sharma, Le Hoang Son, Rohit Sharma, and Korhan Cengiz, editors, *Micro-Electronics and Telecommunication Engineering*, pages 367–388, Singapore. Springer Singapore.

[42] Anna M. Ritcey and Ermanno Borra. (2010). Magnetically deformable liquid mirrors from surface films of silver nanoparticles. *ChemPhysChem*, 11(5):981–986.

[43] Nic Roberts, Noushin Raeisi Kheirabadi, Michail-Antisthenis Tsompanas, Alessandro Chiolerio, Marco Crepaldi, and Andrew Adamatzky. (2024). Logical circuits in colloids. *Royal Society Open*, Article in press.

[44] N.A. Rynin. (1971). K.e. tsiolkovskii life, writings, and rockets. *Interplanetary flight and communication*, III(7):NASA TT F–640.

[45] G. Schubert and K.M. Soderlund. (2011). Planetary magnetic fields: Observations and models. *Physics of the Earth and Planetary Interiors*, 187(3):92–108. Special Issue: Planetary Magnetism, Dynamo and Dynamics.

[46] Uddalok Sen and Souvick Chatterjee. (2019). *Ferrofluids for Propulsion*, pages 287–301. Springer Singapore, Singapore.

[47] Asif A. Siddiqi. (2018). *Beyond Earth - A CHRONICLE OF DEEP SPACE EXPLORATION, 1958–2016*. Office of Communications, NASA History Division, Washington, DC 20546.

[48] Pedro Simplício, Andrés Marcos, Eric Joffre, Mattia Zamaro, and Nuno Silva. (2018). Review of guidance techniques for landing on small bodies. *Progress in Aerospace Sciences*, 103:69–83.

[49] David S. Stevenson. (2016). *The Gas Giants*, pages 273–310. Springer International Publishing, Cham.

[50] Arwen I. I. Tyler, Hanna M. G. Barriga, Edward S. Parsons, Nicola L. C. McCarthy, Oscar Ces, Robert V. Law, John M. Seddon, and Nicholas J. Brooks. (2015). Electrostatic swelling of bicontinuous cubic lipid phases. *Soft Matter*, 11:3279–3286.

[51] Pietro Ubertini, Neil Gehrels, Ian Corbett, Paolo de Bernardis, Marcos Machado, Matt Griffin, Michael Hauser, Ravinder K. Manchanda, Nobuyuki Kawai, Shuang-Nan Zhang, and Mikhail Pavlinsky. (2012). Future of space astronomy: A global road map for the next decades. *Advances in Space Research*, 50(1):1–55.

[52] P. Weissman. (1990). The oort cloud. *Nature*, 344:825–830.

[53] J. Judson Wynne, John E. Mylroie, Timothy N. Titus, Michael J. Malaska, Debra L. Buczkowski, Peter B. Buhler, Paul K. Byrne, Glen E. Cushing, Ashley Gerard Davies, Amos Frumkin, Candice Hansen-Koharcheck, Victoria Hiatt, Jason D. Hofgartner, Trudi Hoogenboom, Ulyana Horodyskyj, Kynan Hughson, Laura Kerber, Margaret Landis, Erin J. Leonard, Elodie Lesage, Alice Lucchetti, Matteo Massironi, Karl L. Mitchell, Luca Penasa, Cynthia B. Phillips, Riccardo Pozzobon, Jani Radebaugh, Francesco Sauro, Robert V. Wagner, and Thomas R. Watters. (2022). Planetary caves: A solar system view of processes and products. *Journal of Geophysical Research: Planets*, 127(11):e2022JE007303. e2022JE007303 2022JE007303.

[54] Xiaofeng Yang, Yishou Wang, and Xinlin Qing. (2018). A flexible capacitive pressure sensor based on ionic liquid. *Sensors*, 18(7).

[55] Maurice Yolles. (2022). Consciousness, sapience and sentience—a metacybernetic view. *Systems*, 10(6).

[56] Ge Zhang, Sungyun Yang, Jing Fan Yang, David Gonzalez-Medrano, Marc Z. Miskin, Volodymyr B. Koman, Yuwen Zeng, Sylvia Xin Li, Matthias Kuehne, Albert Tianxiang Liu, Allan M. Brooks, Mahesh Kumar, and Michael S. Strano. (2024). High energy density picoliter-scale zinc-air microbatteries for colloidal robotics. *Science Robotics*, 9(93):eade4642.

[57] Yongchang Zhang, Pengchun Li, Jiale Quan, Longqiu Li, Guangyu Zhang, and Dekai Zhou. (2023). Progress, challenges, and prospects of soft robotics for space applications. *Advanced Intelligent Systems*, 5(3):2200071.

[58] Xiaoping Zhou, Yi He, and Jie Zeng. (jan 2019). Liquid metal antenna-based pressure sensor. *Smart Materials and Structures*, 28(2):025019.

[59] Xuyan Zhou, Renchang Zhang, Luojia Li, Lunjia Zhang, Bingxin Liu, Zhongshan Deng, Lejing Wang, and Lin Gui. (2019). A liquid metal based capacitive soft pressure microsensor. *Lab Chip*, 19:807–814.